Pl
THE CLUETRAIN MANIFESTO

"One of the best, most eye-opening books I ever read about marketing . . . *The Cluetrain Manifesto* is as potent and relevant now as it was when came out."
—*The Miami Herald*

"The pretentious, strident, and absolutely brilliant creation of four marketing gurus who have renounced marketing-as-usual."
—*Wall Street Journal*

"A book written early enough to not even contain the word 'blog,' but more relevant now than ever."
—*Multichannel Merchant*

"The reason [this book] is still so attractive for business people is that the four authors are, primarily, tech guys . . . so their thoughts are pure, focused, and very different from business-oriented authors."
—*The Gazette* (Montreal)

"Almost ten years ago the seminal book, *The Cluetrain Manifesto*, set out to examine the challenges to business that the Internet posed . . . Well into the first decade of the brave new twenty-first century, it is clear that the changes these prophets spoke of are irreversible."
—*The Star* (South Africa)

"While others work on turning the Internet into the perfect medium for reaching traditional business goals, these four Net-philes hope cyberspace will give commerce a 'human voice.'"
—*Harvard Business Review*

"At its base, this is a rethinking of management and a rabble-rousing invitation to move way outside the box."
—*AudioFile*

"A different brand of business book, thank goodness: saucy, heartfelt, and warmly appealing in its faith in the commonwealth."
—*Kirkus*

"The most important business book since *In Search of Excellence*. Get a clue. Read the book."
—*Information Week*"

You must read *The Cluetrain Manifesto*. So: read it, inhale it. If it pisses you off . . . GREAT!"
—Tom Peters

"Love it or hate it, no one with a stake in the online scene can afford to ignore what this book is saying."
—Michael Wolff

"If you don't think you need this book to better understand your market, that's your second mistake!"
—Seth Godin

"I've seen the future of business, and it's *The Cluetrain Manifesto*."
—Thomas Petzinger, Jr.,
author of *The New Pioneers*

TENTH ANNIVERSARY EDITION

THE CLUETRAIN MANIFESTO

RICK LEVINE

+

CHRISTOPHER LOCKE

+

DOC SEARLS

+

DAVID WEINBERGER

BASIC
BOOKS

A Member of the Perseus Books Group
New York

Copyright © 2000, 2001, 2009 by Fredrick Levine, Christopher Locke, David Searls, and David Weinberger

Tenth Anniversary Edition published in 2009 by Basic Books,
A Member of the Perseus Books Group
Tenth Anniversary Paperback Edition published in 2011 by Basic Books

Previously published by Perseus Publishing

Books published by Basic Books are available at special discounts for bulk purchases in the United States by corporations, institutions, and other organizations. For more information, please contact the Special Markets Department at the Perseus Books Group, 2300 Chestnut Street, Suite 200, Philadelphia, PA 19103, or call (800) 810-4145, ext. 5000, or e-mail special.markets@perseusbooks.com.

Text Design by Rose Thaynor

The Library of Congress has catalogued the hardcover as follows:
Levine, Rick.
    The cluetrain manifesto: The end of business as usual / Rick Levine.
    p. cm.
    Includes index.
    ISBN 978-0-465-01865-9 (alk. paper)
    1. Electronic commerce—Social aspects. 2. Customer relations—Technological innovations. 3. Internet marketing. 4. Intranets (Computer networks). 5. Corporate culture. 6. Information superhighway—Economic aspects. I. Title.
HF5548.32 .C56 2009
303.48'33 dc—22

                                                                                    2009903733

Paperback ISBN: 978-0-465-02409-4
e-book ISBN: 978-0-465-00414-0

10 9 8 7 6 5 4 3 2 1

CONTENTS

CHRISTOPHER LOCKE

Ancient markets were full of the sound of life: conversation. They dealt in craft goods that bore the marks of the people who made them. Then mass production led to mass marketing and mass media: interchangeable workers, products, and consumers, along with the hierarchical bureaucracies needed to command and control them. But the Internet—unmanaged and full of the sound of the human voice—is proving that the Industrial Age is nothing but an interruption as the conversations resume, this time on a global scale.

DAVID WEINBERGER

Our culture has greeted the Web with such enthusiasm, even before we've understood what exactly it's for, because we believe it's returning something we miss deeply: our voices. We are now ending the Faustian bargain according to which we gave up much of our individuality at work in return for the illusion of living in a manageable, safe world. As we emerge into the new understanding, it is being revealed that this cultural change is in fact spiritual.

RICK LEVINE

Voice—the authentic expression of the individual present in the craft-work of our hands as well as in our words—is in resurgence on the Internet. In fact, the Internet is a conversation carried on in a variety of formats—Web pages, e-mail, discussion groups, mailing lists—that bring new possibilities to human relations. Business-as-usual isn't happy about this, because conversations are unpredictable, messy, and uncontrollable. But there's no silencing our human voices. Wise companies will learn how to enter the conversation.

DOC SEARLS AND DAVID WEINBERGER

The mass production of the industrial world led companies to engage
in mass marketing, delivering "messages" to undifferentiated hordes
who didn't want to receive them. Now the Web is enabling the market to converse again, as people tell one another the truth about
products and companies and their own desires—learning faster than
business. Companies have to figure out how to enter this global conversation rather than relying on the old marketing techniques of public relations, marketing communications, advertising, and other forms
of propaganda. We, the market, don't want messages at all, we want
to speak with your business in a human voice.

DAVID WEINBERGER

Like their counterparts in the marketplace, intranetworked employees
are learning to speak in their own voice, ignoring the org charts, and
telling the truth to one another—and to their customers. The direct
connection, worker to worker, enabled by intranets is undermining
the old management pyramid, turning the walls of Fort Business into
conversations. From the bottom up, businesses are beginning to
accept openness, decentralization, fallibility, messy context-rich information, stories, and the sound of the authentic voices of individuals.

CHRISTOPHER LOCKE AND DAVID WEINBERGER

We're living through a change of historic proportions, so there are no
easy answers. In fact, there aren't even any easy questions any more.
The ones most commonly asked—in magazine articles and political
speeches—are disguised attempts to distract us from the truly deep
changes that are occurring. We'd do far better to learn to listen to the
questions coming from our hearts.

TAKE A POT OF WATER THAT'S JUST ABOVE THE FREEZING MARK. Now, crank up the heat and wait. Temperature rises. Wait some more. Go all the way to 211 degrees Fahrenheit and nothing looks much different. But then, turn it up one more tiny degree, and wham! The pot becomes a roiling, steamy cauldron.

Don't look now, but you're holding such a catalyst in your hands. THE CLUETRAIN MANIFESTO is about to drive business to a full boil.

Let me tell you how it took me to the tipping point. Not long ago I was sitting in the Hotel Nikko in San Francisco on a reporting mission for "The Front Lines," a weekly column I spent four years writing for THE WALL STREET JOURNAL. Between interviews, I was checking e-mail from my readers. (The Internet puts me in touch with thousands of them who act as my scouts.) On this particular day, one of my correspondents urged me to check out a new site at www.cluetrain.com.

I was dumbstruck. There, in a few pages, I read a startlingly concise summary of everything I'd seen in twenty-one years as a reporter, editor, bureau chief, and columnist for my newspaper. The idea that business, at bottom, is fundamentally human. That engineering remains second-rate without aesthetics. That natural, human conversation is the true language of commerce. That corporations work best when the people on the inside have the fullest contact possible with the people on the outside.

And most importantly, that however ancient, timeless, and true, these principles are just now resurging across the business world. The triggering event, of course, is the advent of a global communication system that restores the banter of the bazaar, that tears down power structures and senseless bureaucracies, that puts everyone in touch with everyone.

Scrolling through the hundreds of signatories who had endorsed the manifesto, I realized this if nothing else: The newspaper gods had just blessed me with one of my favorite columns ever, enabling me to articulate much I knew to be true but never previously had the words to say.

Because "The Front Lines" was usually a narrative tale, I bored into the manifesto's origins. Befitting its message, the document, I learned, was born in an extended electronic conversation among four Internet denizens spread from coast to coast. The authors were not the ultra-hip, just-outta-college webheads I had imagined. One was Rick Levine, a Boulder-based engineer for the giant Sun Microsystems. Another was a Boulder consultant named Christopher Locke, late of such hoary outfits as IBM, MCI, and Carnegie Mellon. There was a well-known Silicon Valley publicist named Doc Searls and a longtime high-tech marketer from Boston whose name, David Weinberger, I recognized from his commentaries on National Public Radio's "All Things Considered."

They were, in short, fixtures of the high-tech establishment—but being establishment made their renunciation of business-as-usual all the more powerful.

The manifesto URL leaped between cubicles like mononucleosis through a co-ed dorm. Some readers found it pretentious, bordering on smug. (To those of delicate sensibility, it was.) Some found it nihilistic. (It wasn't.) But all found it arresting and impossible to ignore. The manifesto became a kind of user's guide to the Internet economy—a world of new online communities; of self-organizing corporate employees; of Linux and other "open source" movements that seem to erupt from thin air.

So now, for anyone who missed it the first time and for everyone else who wants more, we have THE CLUETRAIN MANIFESTO, one of the first books written as sequel to a Web site.

I look at a huge number of business books. I actually read some of them and have published reviews on more than my share. I'll mention a few ways THE CLUETRAIN MANIFESTO is like no other.

First, this is no feel-good book. Though the broad theme is overwhelmingly optimistic, the details will make you squirm. This is an obituary for business-as-usual. It shows how your Web strategy may be minutes from obsolescence. It reveals how the Internet has made your entry-level employees as powerful as your senior vice president of marketing. Recall what THE JUNGLE did to meat packing, what SILENT SPRING did to chemicals, what UNSAFE AT ANY SPEED did to Detroit. That's the spirit with which THE CLUETRAIN MANIFESTO takes on the arrogance of corporate e-commerce. (Notably, some of the best material comes from the authors' own experiences within big companies, and they name names.)

Second, this is not a how-to book, unless you need a remedial lesson in being human. For all their righteous self-assuredness about the Internet revolution, these authors don't presume to tell you how to run your business or your career. One-size-fits-all "programs" and "methodologies" are just ways for consultants to gouge clients and book buyers. Instead, this book simply describes business *as it really is* and *as it's really becoming.* You'll come away from these pages with a new set of eyes for redirecting your career or rehabilitating your company according to its own unique circumstances.

Third, this book is not boring. The whole message here, after all, involves speaking with a human voice. That means stories instead of lectures, humor instead of hubris, description instead of PowerPoint pie charts. (Imagine IN SEARCH OF EXCELLENCE crossed with FEAR AND LOATHING IN LAS VEGAS.) When was the last time you laughed out loud reading a business book?

And why not laughter? It's one of the signature melodies of human conversation. This book shows how conversation forms the basis of business, how business lost that voice for a while, and how that language is returning to business thanks to a technology that inspires, and in many cases demands, that we speak from the heart.

To rip off what rock critic Jon Landau once said about Bruce Springsteen: I've seen the future of business, and it's THE CLUETRAIN MANIFESTO. At first you may be tempted to hide this book inside the dust jacket for customers.com or something equally conventional. But in time you'll see the

book spreading. It will become acceptable, if never entirely accepted. It will certainly become essential. Why am I so sure? Because like nothing else out there,
it shows us how to grasp the human side of business and technology, and being
human, try as we might, is the only fate from which we can never escape.

—**Thomas Petzinger, Jr.**
The Wall Street Journal

People of Earth...

A POWERFUL GLOBAL CONVERSATION HAS BEGUN.
Through the Internet, people are discovering and inventing new ways to share
relevant knowledge with blinding speed. As a direct result, markets are getting
smarter—and getting smarter faster than most companies.

These markets are conversations. Their members communicate in language
that is natural, open, honest, direct, funny, and often shocking. Whether explaining or complaining, joking or serious, the human voice is unmistakably genuine.
It can't be faked.

Most corporations, on the other hand, only know how to talk in the soothing,
humorless monotone of the mission statement, marketing brochure, and your-
call-is-important-to-us busy signal. Same old tone, same old lies. No wonder networked markets have no respect for companies unable or unwilling to speak as
they do.

But learning to speak in a human voice is not some trick, nor will corporations convince us they are human with lip service about "listening to customers."

They will only sound human when they empower real human beings to speak on their behalf.

While many such people already work for companies today, most companies ignore their ability to deliver genuine knowledge, opting instead to crank out sterile happytalk that insults the intelligence of markets literally too smart to buy it.

However, employees are getting hyperlinked even as markets are. Companies need to listen carefully to both. Mostly, they need to get out of the way so intranetworked employees can converse directly with internetworked markets.

Corporate firewalls have kept smart employees in and smart markets out. It's going to cause real pain to tear those walls down. But the result will be a new kind of conversation. And it will be the most exciting conversation business has ever engaged in.

95 Theses

1 Markets are conversations.

2 Markets consist of human beings, not demographic sectors.

3 Conversations among human beings *sound* human. They are conducted in a human voice.

4 Whether delivering information, opinions, perspectives, dissenting arguments, or humorous asides, the human voice is typically open, natural, uncontrived.

5 People recognize each other as such from the sound of this voice.

6 The Internet is enabling conversations among human beings that were simply not possible in the era of mass media.

7 Hyperlinks subvert hierarchy.

8 In both *inte*rnetworked markets and among *intra*networked employees, people are speaking to each other in a powerful new way.

9 These networked conversations are enabling powerful new forms of social organization and knowledge exchange to emerge.

10 As a result, markets are getting smarter, more informed, more organized. Participation in a networked market changes people fundamentally.

11 People in networked markets have figured out that they get far better information and support from one another than from vendors. So much for corporate rhetoric about adding value to commoditized products.

12 There are no secrets. The networked market knows more than companies do about their own products. And whether the news is good or bad, they tell everyone.

13 What's happening to markets is also happening among employees. A metaphysical construct called "The Company" is the only thing standing between the two.

14 Corporations do not speak in the same voice as these new networked conversations. To their intended online audiences, companies sound hollow, flat, literally inhuman.

15 In just a few more years, the current homogenized "voice" of business— the sound of mission statements and brochures—will seem as contrived and artificial as the language of the 18th-century French court.

16 Already, companies that speak in the language of the pitch, the dog-and-pony show, are no longer speaking to anyone.

17 Companies that assume online markets are the same markets that used to watch their ads on television are kidding themselves.

18 Companies that don't realize their markets are now networked person-to-person, getting smarter as a result and deeply joined in conversation, are missing their best opportunity.

19 Companies can now communicate with their markets directly. If they blow it, it could be their last chance.

20 Companies need to realize their markets are often laughing. At them.

21 Companies need to lighten up and take themselves less seriously. They need to get a sense of humor.

22 Getting a sense of humor does not mean putting some jokes on the corporate web site. Rather, it requires big values, a little humility, straight talk, and a genuine point of view.

23 Companies attempting to "position" themselves need to *take* a position. Optimally, it should relate to something their market actually cares about.

24 Bombastic boasts—"We are positioned to become the preeminent provider of XYZ"—do not constitute a position.

25 Companies need to come down from their Ivory Towers and talk to the people with whom they hope to create relationships.

26 Public Relations does not relate to the public. Companies are deeply afraid of their markets.

27 By speaking in language that is distant, uninviting, arrogant, they build walls to keep markets at bay.

28 Most marketing programs are based on the fear that the market might see what's really going on inside the company.

29 Elvis said it best: "We can't go on together with suspicious minds."

30 Brand loyalty is the corporate version of going steady, but the breakup is inevitable—and coming fast. Because they are networked, smart markets are able to renegotiate relationships with blinding speed.

31 Networked markets can change suppliers overnight. Networked knowledge workers can change employers over lunch. Your own "downsizing initiatives" taught us to ask the question: "Loyalty? What's that?"

32 Smart markets will find suppliers who speak their own language.

33 Learning to speak with a human voice is not a parlor trick. It can't be "picked up" at some tony conference.

34 To speak with a human voice, companies must share the concerns of their communities.

35 But first, they must belong to a community.

36 Companies must ask themselves where their corporate cultures end.

37 If their cultures end before the community begins, they will have no market.

38 Human communities are based on discourse—on human speech about human concerns.

39 The community of discourse *is* the market.

40 Companies that do not belong to a community of discourse will die.

41 Companies make a religion of security, but this is largely a red herring. Most are protecting less against competitors than against their own market and workforce.

42 As with networked markets, people are also talking to each other directly *inside* the company—and not just about rules and regulations, boardroom directives, bottom lines.

43 Such conversations are taking place today on corporate intranets. But only when the conditions are right.

44 Companies typically install intranets top-down to distribute HR policies and other corporate information that workers are doing their best to ignore.

45 Intranets naturally tend to route around boredom. The best are built bottom-up by engaged individuals cooperating to construct something far more valuable: an intranetworked corporate conversation.

46 A healthy intranet *organizes* workers in many meanings of the word. Its effect is more radical than the agenda of any union.

47 While this scares companies witless, they also depend heavily on open intranets to generate and share critical knowledge. They need to resist the urge to "improve" or control these networked conversations.

48 When corporate intranets are not constrained by fear and legalistic rules, the type of conversation they encourage sounds remarkably like the conversation of the networked marketplace.

49 Org charts worked in an older economy where plans could be fully understood from atop steep management pyramids and detailed work orders could be handed down from on high.

50 Today, the org chart is hyperlinked, not hierarchical. Respect for hands-on knowledge wins over respect for abstract authority.

51 Command-and-control management styles both derive from and reinforce bureaucracy, power tripping, and an overall culture of paranoia.

52 Paranoia kills conversation. That's its point. But lack of open conversation kills companies.

53 There are two conversations going on. One inside the company. One with the market.

54 In most cases, neither conversation is going very well. Almost invariably, the cause of failure can be traced to obsolete notions of command and control.

55   As policy, these notions are poisonous. As tools, they are broken. Command and control are met with hostility by intranetworked knowledge workers and generate distrust in internetworked markets.

56   These two conversations want to talk to *each other*. They are speaking the same language. They recognize each other's voices.

57   Smart companies will get out of the way and help the inevitable to happen sooner.

58   If willingness to get out of the way is taken as a measure of IQ, then very few companies have yet wised up.

59   However subliminally at the moment, millions of people now online perceive companies as little more than quaint legal fictions that are actively preventing these conversations from intersecting.

60   This is suicidal. Markets *want* to talk to companies.

61   Sadly, the part of the company a networked market wants to talk to is usually hidden behind a smokescreen of hucksterism, of language that rings false—and often is.

62   Markets do not want to talk to flacks and hucksters. They want to participate in the conversations going on behind the corporate firewall.

63   De-cloaking, getting personal: We *are* those markets. We want to talk to *you*.

64   We want access to your corporate information, to your plans and strategies, your best thinking, your genuine knowledge. We will not settle for the four-color brochure, for web sites chock-a-block with eye candy but lacking any substance.

65   We're also the workers who make your companies go. We want to talk to customers directly in our own voices, not in platitudes written into a script.

66   As markets, as workers, both of us are sick to death of getting our information by remote control. Why do we need faceless annual reports and third-hand market research studies to introduce us to each other?

67   As markets, as workers, we wonder why you're not listening. You seem to be speaking a different language.

68   The inflated self-important jargon you sling around—in the press, at your conferences—what's that got to do with us?

69   Maybe you're impressing your investors. Maybe you're impressing Wall Street. You're not impressing us.

70 If you don't impress us, your investors are going to take a bath. Don't they understand this? If they did, they wouldn't *let* you talk that way.

71 Your tired notions of "the market" make our eyes glaze over. We don't recognize ourselves in your projections—perhaps because we know we're already elsewhere.

72 We like this new marketplace much better. In fact, we are creating it.

73 You're invited, but it's our world. Take your shoes off at the door. If you want to barter with us, get down off that camel!

74 We are immune to advertising. Just forget it.

75 If you want us to talk to you, tell us something. Make it something interesting for a change.

76 We've got some ideas for you too: some new tools we need, some better service. Stuff we'd be willing to pay for. Got a minute?

77 You're too busy "doing business" to answer our e-mail? Oh gosh, sorry, gee, we'll come back later. Maybe.

78 You want us to pay? We want you to pay attention.

79 We want you to drop your trip, come out of your neurotic self-involvement, join the party.

80 Don't worry, you can still make money. That is, as long as it's not the only thing on your mind.

81 Have you noticed that, in itself, money is kind of one-dimensional and boring? What else can we talk about?

82 Your product broke. Why? We'd like to ask the guy who made it. Your corporate strategy makes no sense. We'd like to have a chat with your CEO. What do you mean she's not in?

83 We want you to take 50 million of us as seriously as you take one reporter from THE WALL STREET JOURNAL.

84 We know some people from your company. They're pretty cool online. Do you have any more like that you're hiding? Can they come out and play?

85 When we have questions we turn to each other for answers. If you didn't have such a tight rein on "your people" maybe they'd be among the people we'd turn to.

86 When we're not busy being your "target market," many of us *are* your people. We'd rather be talking to friends online than watching the clock. That would get your name around better than your entire million-dollar web site. But you tell us speaking to the market is Marketing's job.

87 We'd like it if you got what's going on here. That'd be real nice. But it would be a big mistake to think we're holding our breath.

88 We have better things to do than worry about whether you'll change in time to get our business. Business is only a part of our lives. It seems to be all of yours. Think about it: Who needs whom?

89 We have real power and we know it. If you don't quite see the light, some other outfit will come along that's more attentive, more interesting more fun to play with.

90 Even at its worst, our newfound conversation is more interesting than most trade shows, more entertaining than any TV sitcom, and certainly more true-to-life than the corporate web sites we've been seeing.

91 Our allegiance is to ourselves—our friends, our new allies and acquaintances, even our sparring partners. Companies that have no part in this world also have no future.

92 Companies are spending billions of dollars on Y2K. Why can't they hear this market timebomb ticking? The stakes are even higher.

93 We're both inside companies and outside them. The boundaries that separate our conversations look like the Berlin Wall today, but they're really just an annoyance. We know they're coming down. We're going to work from both sides to *take* them down.

94 To traditional corporations, networked conversations may appear confused, may sound confusing. But we are organizing faster than they are. We have better tools, more new ideas, no rules to slow us down.

95 We are waking up and linking to each other. We are watching. But we are not waiting.

The Elevator Rap

WHEN

(Inter) networked
Markets

meet

(Intra) networked
Workers

**The connectedness of the Web is transforming what's inside
and outside your business—your market and your employees.**

Through the Internet, the
people in your markets
are discovering and
inventing new ways to
converse. They're talking
about your business.
They're telling one
another the truth,
in very human voices.

Intranets are enabling
your best people to
hyperlink themselves
together, outside the org
chart. They're incredibly
productive and innovative.
They're telling one
another the truth,
in very human voices.

**There's a new conversation between and among your market
and your workers. It's making them smarter and it's enabling them
to discover their human voices.**

You have two choices. You can continue to lock yourself behind
facile corporate words and happytalk brochures.

Or you can join the conversation.

INTRODUCTION

WHAT IF THE REAL ATTRACTION OF THE INTERNET IS NOT its cutting-edge bells and whistles, its jazzy interface, or any of the advanced technology that underlies its pipes and wires? What if, instead, the attraction is an atavistic throwback to the prehistoric human fascination with telling tales? Five thousand years ago, the marketplace was the hub of civilization, a place to which traders returned from remote lands with exotic spices, silks, monkeys, parrots, jewels—and fabulous stories.

In many ways, the Internet more resembles an ancient bazaar than it fits the business models companies try to impose upon it. Millions have flocked to the Net in an incredibly short time, not because it was user-friendly—it wasn't—but because it seemed to offer some intangible quality long missing in action from modern life. In sharp contrast to the alienation wrought by homogenized broadcast media, sterilized mass "culture," and the enforced anonymity of bureaucratic organizations, the Internet connected people to each other and provided a space in which the human voice would be rapidly rediscovered.

Though corporations insist on seeing it as one, the new marketplace is not necessarily a market at all. To its inhabitants, it is primarily a place in which all participants are audience to each other. The entertainment is not packaged; it is intrinsic. Unlike the lockstep conformity imposed by television, advertising, and corporate propaganda, the Net has given new legitimacy—and free rein—to play. Many of those drawn into this world find themselves exploring a freedom never before imagined: to indulge their curiosity, to debate, to disagree, to laugh at themselves, to compare visions, to learn, to create new art, new knowledge.

Because the Internet is so technically efficient, it has also been adopted by companies seeking to become more productive. They too are hungry for knowledge, for the intellectual capital that has become more valuable than bricks and mortar or any tangible asset. What they didn't count on were the other effects of Web technology. Hypertext is inherently nonhierarchical and antibureaucratic. It does not reinforce loyalty and obedience; it encourages idle speculation and loose talk. It encourages stories.

These new conversations online—whether on the wild and wooly Internet or on (slightly) more sedate corporate intranets—are generating new ways of looking at problems. They are spawning new perspectives, new tools, and a new kind of intellectual bravery more comfortable with risk than with regulation. The result is not just new things learned but a vastly enhanced ability to learn things. And the pace of this learning is accelerating. In the networked marketplace it is reflected in the joy of play. On company intranets it is reflected in the joy of knowledge. But it's getting difficult to tell the two apart. Employees go home and get online. They bring new attitudes back to work the next day. Enthusiastic surfers get hired and bring strange new views into corporations that, until now, have successfully protected themselves from everything else. The World Wide Web reinforces freedom. The Internet routes around obstacles. The confluence of these conversations is not only inevitable, it has largely already occurred.

Many companies fear these changes, seeing in them only a devastating loss of control. But control is a losing game in a global marketplace where the range of customer choice is already staggering and a suicidal game for companies that must come up with the knowledge necessary to create those market choices.

While command and control may have reached a cul-de-sac, the intersection of the market conversation with the conversation of the corporate workforce hardly signals the end of commerce. Instead, this convergence promises a vibrant renewal in which commerce becomes far more naturally integrated into the life of individuals and communities.

THIS BOOK TELLS A STORY. FOUR TIMES. MANY TIMES.
It is the story of how these things have happened—and some powerful hints about what could happen from here on out.

The End?
Of Business?
As Usual?

DAVID WEINBERGER

WHEN THE CLUETRAIN MANIFESTO CAME OUT TEN YEARS AGO, ITS subtitle made a prediction or a threat, depending on which side of the corporate desk your chair was placed on: THE END OF BUSINESS AS USUAL. A decade later, not that much seems to have changed, except for our book's subtitle. Business has had its ups and downs, but that's, well, usual for business.

Did the original edition of CLUETRAIN commit the Fallacy of Hyperbolic Subtitle?

It depends which of the four of us you ask. For, in this tenth anniversary edition—which, because it adds about half again to the original text, should perhaps be called a tenth anniversary *addition*—we are in a peculiar position. The four authors disagree about the extent to which we got it right, but the three guest contributors are less hesitant. Jake McKee (Lego), J. P. Rangaswami (British Telecom), and Dan Gillmor (the highly esteemed journalist who has led the way with citizen journalism) think CLUETRAIN definitely was on to something.

On to what? Five years before the idea of Web 2.0 was born, CLUETRAIN said that the Web was not what business and the media were insisting it was. Those institutions looked into the deep pond of the Web and saw their own reflections. To business, the Web looked like nothing but a chance to advertise, to market, and to sell stuff. To the media, the Web looked like a publishing medium. Of course the Web is both those things, but most of all (CLUETRAIN said), it is a place where we humans get to talk with one another in our own voices about what matters to us. For you cannot explain the world's rapid embrace of this odd new technology by saying that we humans were so inspired by the prospect of paperless catalog shopping. Nor can you attribute it to an unprecedented global desire to become research librarians. No, said CLUETRAIN, the Web is touching our most ancient of needs: to connect.

And to talk. Doc Searls's phrase, which predated CLUETRAIN, stuck: markets are conversations. They were conversations before the Industrial Revolution (woven through CLUETRAIN is a historical thread) and now, thanks to the Net, they were again. Most of all, markets are conversations among customers, and marketers better think twice before they stick their noses in. We treasure our conversations most of all because they are *ours*, the way marketing-speak never is.

While CLUETRAIN was pitched as a business book, and the best-known phrase from it talked about markets, we never thought it was only about business. Business is just the best example we could find. Markets are conversations, but so are businesses, governments, schools, and the Net. Of course, they are also more than that, but noticing and valuing the conversational side of each of those institutions still draws our attention to the aspects that are more unsettled, improvisational, passionate, and human.

So, were we right in our implicit prediction that business as usual was about to end? Not exactly. Huge corporations still stalk the earth. We still report to hierarchical structures that cut us paychecks in exchange for obedience. Our bosses still trend more toward a-holism than do our coworkers. Television commercials have expanded from the interstices to overlaying the very programs we're watching. In fact, the Web has given marketers opportunities to betray us in more inventive ways.

All four of the authors cop to that. Yup. But a quick and obvious "we were wrong" actually misses the ways we were right. The problem is, our culture has absorbed so many of these changes that it's easy to miss just how weird the new normal is. For example:

There are tens of millions of active blogs.

Social networking sites—Facebook, MySpace, and let's count Twitter here too—have created a new platform for one of our most basic relationships: friendship.

Online retailers routinely let their customers review the products retailers are trying to sell. Bad reviews stand next to good ones, untouched.

Customers will leave a site that does not let them find exactly what they want without having to wade through pages that flog products the company wants to sell them.

Users post in public places the pages that they want to bookmark, and these social bookmarks are then used as a new system of categorization and classification of other people's "content," including businesses'.

A company's site has become one of the last places customers go when they have questions. The customers are happily, yes, conversing on blogs, review sites, and forums.

The recording industry has been subverted by an ethos of sharing.

Newspapers are on the verge of collapse as their functionality has been taken over by more efficient mechanisms: Distribution by anything with a screen, coverage by syndicates such as AP and Reuters, commentary by everyone with an opinion, editorial judgment by everyone who can make a recommendation via email or a Website, and classified ads by a guy named Craig. Of course, it's uncertain that we're going to be able to replace all of what we're losing, but for newspapers, it's certainly not business as usual.

Some companies are far more transparent with their customers and employees than they were. So are some politicians.

Working from home has become commonplace for many people.

Leaders who lead as if they are the highest incarnation of Business Man look increasingly foolish. We've learned that some of the largest projects cannot be accomplished so long as leaders get in the way.

Scholars and scientists are coming to the conclusion that the old system of scholarly publishing is a crime against their vocation. They are inventing ways to share their work more widely, earlier, and more intimately.

Customers and citizens always expect more information. No matter how much they have, they expect links to another set of pages.

The mean time before frustration with companies that are unresponsive to customer inquiries has dropped from weeks to minutes.

Companies and occasional industries have realized that they do better if they give information away instead of treating it as an asset to be guarded. Case in point: The NEW YORK TIMES finally made its archives available, in part because charging for access kept its wealth of information from showing up in search results, hurting the newspaper's prestige and its online ad revenues.

Instant messaging, which companies at first thought was for kids and gossip, now has become a standard part of many companies' internal and external infrastructures. Next up: social networks and Twitter.

Companies routinely run blogs where employees (including senior managers sometimes) talk like actual people.

On these blogs, and through other means, companies are far more likely to disclose possible future directions than they were before.

While some products of course remain scarce, the information about products is super-abundant. Company strategies that try to control the stream of information routinely now simply make the company look like a scaredy-cat.

Google has become a verb.

The small actions of many individuals can be aggregated in "crowd sourcing" applications that do everything from notify a city of burned-out streetlights to document acts of violence during the disturbances in Kenya in January 2008.

A presidential candidate won an election in part because he successfully combined bottom-up organization with sufficient top-down coordination.

We built the world's greatest encyclopedia in our spare time.

Creating, watching, rating, recommending, and talking about homemade videos is eating into the time we are supposed to spend every day narcotized in front of the TV.

With these boons has come an expanding range of online horror: sites that appeal to the worst in us, tools for plotting attacks that intend to terrorize us, online bullying, global botnets that rip us off, calculated invasions of our privacy, and the unimagined depravities sure to turn up tomorrow. But the question at the moment isn't whether the Web is good or bad. (The answer to that is yes.) The question is how hyperbolic was CLUETRAIN's original subtitle.

The list of individual changes, large and small, points to four profound shifts in what business as usual used to take for granted.

First, companies used to be in an us-them relationship with their customers. The business had sharp edges and sometimes sharp elbows. Now the lines are smudged. Sure, there's an important financial and legal difference between an employee and a customer, and there always (always?) will be. But business

What does "Cluetrain" mean?

Only after the first edition of the original *Cluetrain* book was published did we realize that we had neglected to explain where the title came from. We weren't trying to be mysterious. In fact, the Cluetrain.com site had explained it from the beginning. We just forgot to put it in the book. During one of the long phone conversations that gave rise to the book, Doc was reminded of a Silicon Valley company about which a friend had said, "The clue train stopped there four times a day, and they never took delivery." We all laughed. Ten minutes later, Doc interjected into the phone conversation that he had just registered cluetrain.com.

processes now routinely transgress those lines, from inviting customers into the product planning process to participating in the online reviews.

Second, customers are so empowered that they don't feel especially empowered. The new normal is that we expect businesses to listen to us. The companies that don't are now perceived quite clearly as dinosaurs.

Third, we've broken the back of the old realism. It turns out that we can do more than we thought. It's hard to grasp what's realistic because the scale of connections is unthinkable. It's like thinking that natural selection cannot account for eyeballs because it's so hard to imagine what 4.5 billion years of life could do.

Fourth, the most important realistic idea that lies broken at our feet is that ultimately we're individuals out for our own good. There are now too many spectacularly successful projects to hold on to that characterization of human beings. We are selfish, sure, but we are also generous. We create sites for free, and then we'll put in links to entice people to go somewhere else. We build amazing projects that the old realism would have dismissed as mere subtitle hyperbolism.

Lots of the old ways of doing and thinking about business remain. Of course. But we're not done yet, we people of earth. A powerful global conversation has begun...

Markets Are Relationships

DOC SEARLS

WHEN ALL YOU'VE GOT IS A HAMMER, BAD SERVICE LOOKS LIKE A NAIL.

In August 2007, Mona Shaw took a hammer to her local Comcast office. Literally. First, BAM! She blasted the customer service rep's keyboard. Then BOOM! She took out a monitor. Then POW! She destroyed a phone. People screamed and ran. When the cops showed up, WHACK! She hammered the phone, one more time. Up to this point, there was nothing exceptional about Mrs. Shaw. She was a retired nurse. A grandmother. She took in stray dogs. She went to church every Sunday, and was the secretary for both her local AARP and a square dance club. What made her snap was something even less exceptional: awful customer service.

According to Mona and her husband Don,[1] it all began when they signed up for Comcast's much-promoted Triple Play—a deal that combined telephone, TV, and Internet service. First the Comcast service guy failed to show up on the appointed Monday. When he did finally come, two days later, he didn't finish the job and left tools behind. Meanwhile, Comcast changed the Shaws' phone number, which they had been using for thirty-four years. When Mona called Comcast on a cell phone,[2] she got lost in the company's call center maze. On Friday morning Comcast cut off all three of the Shaw's services. That afternoon Mona and Don went to Comcast's local office and asked to see the manager. A customer service rep told them somebody would be right with them and to sit outside on a bench—in the August heat. Two hours went by before somebody leaned out the door, said the manager had left for the weekend, and thanked the couple for coming. The next Monday, Mona returned with her hammer and got everybody's attention.

At that point in time, THE CLUETRAIN MANIFESTO was more than eight years old. Its subhead—THE END OF BUSINESS AS USUAL—did not apply to Comcast. Or to most forms of "customer service."

Unstarted Business

NOT LONG AFTER THE CLUETRAIN MANIFESTO CAME OUT, I WAS having a beer with Jakob Nielsen, one of the world's top website usability experts and an unusually insightful guy. The subject at hand was CLUETRAIN's success. Why did it so quickly become a best-seller? I had my own theories, but Jakob blew them away with an observation I hadn't heard before and haven't shaken since: "You guys defected from marketing, and sided with markets *against* marketing."

He was right.

When we were starting work on the Cluetrain website in early 1999, the project wasn't on the front burner for any of us. Chris Locke put it there when he sent the rest of us an email with a little graphic that said, "We are not seats or eyeballs or end users or consumers. We are human beings, and our reach exceeds your grasp. Deal with it."

Note the first-person *we*. Not the second-person *you*.

That statement adrenalized us. It also gave us our voice.

Jakob said CLUETRAIN's voice was that of the market, not of marketing— and that this was why CLUETRAIN adrenalized readers too. We spoke to, and for, all human beings who know in their bones that the Net empowers every-body connected through it: buyers as well as sellers, and not just because sell-ers have cool new ways of "serving the customer."

Yet ten years have gone by, and customer reach still does not exceed seller grasp. Not when companies still speak of "acquiring," "owning," "controlling," and "locking in" customers as if they were slaves. Not when "customer sup-port" sends callers down mazes of discouraging choices, and scripted "conversa-tion" is outsourced to weary laborers in far corners of the globe. Not when "relationships" are based on terms-of-service "agreements" that nobody reads and give all advantages to the seller. Not when an old lady can bust up a "ser-vice" office because she can't get any.

While there is much to love about doing business on the Net, there is much more that's still a pain in the ass. Think about it. Do you need to sign an agree-ment to shop at a department store? Do you need to login and enter a pass-word to do business at a dress shop or a shoe store? No. That's because the physical world of business is still more sensible than the virtual one.

But let's be fair. It's still early. E-commerce is just four years older than CLUETRAIN. Real-world business has been around since Ur. There is much catching up to do.

For doing that, I see two ways we can go. One is to wait until "customer care" loses its irony. The other is to take action—to create tools that put cus-tomers in the driver's seat rather than in the back of the bus. This chapter is about the latter choice.

The Assignment

NOT LONG AFTER CLUETRAIN CAME OUT IN EARLY 2000, I FOUND myself on a cross-country flight, sitting beside a Nigerian pastor named Sayo Ajiboye. After we began to talk, it became clear to me that Sayo (pronounced "Shaiyo") was a deeply wise man. Among his accomplishments was translating the highly annotated Thompson Bible into his native Yoruba language: a project that took eight of his thirty-nine years.

I told him that I had been involved in a far more modest book project—THE
CLUETRAIN MANIFESTO—and was traveling the speaking circuit, promoting it.
When Sayo asked me what the book was about, I explained how "markets are
conversations" was the first of our ninety-five theses, and how we had
unpacked it in a chapter by that title. Sayo listened thoughtfully, then came
back with the same response I had heard from other readers in what back then
was still called the Third World: "Markets are conversations" is a pretty smart
thing for well-off guys from the First World to be talking about. But it doesn't
go far enough.

When I asked him why, he told me to imagine we were in a "natural"
marketplace—a real one in, say, an African village where one's "brand" was a
matter of personal reputation, and where nobody ruled customer choices with
a pricing gun. Then he picked up one of those blue airline pillows and told me
to imagine it was a garment, such as a coat, and that I was interested in buying
it. "What's the first thing you would say to the seller?" he asked.

"What does it cost?"

"Yes, you would say that," he replied, meaning that this was typical of a
First World shopper for whom price is the primary concern. Then he asked me
to imagine that a conversation follows between the seller and me—that the
two of us get to know each other a bit and learn from each other. "Now," he
asked, "What happens to the price?"

I said maybe now I'm willing to pay more while the seller is willing to
charge less.

"Why?" Sayo asked.

I didn't have an answer.

"Because you now have a *relationship*," he said.

As we continued talking, it became clear to me that everything that happens
in a marketplace falls into just three categories: *transaction*, *conversation*, and
relationship. In our First World business culture, transaction matters most, con-

versation less, and relationship least. Worse, we conceive and justify everything in transactional terms. Nothing matters more than price and "the bottom line." By looking at markets through the prism of transaction, or even conversation, we miss the importance of relationship. We also don't see how relationship has a value all its own: one that transcends, even as it improves, the other two.

Consider your relationship with friends and family, Sayo said. The value system there is based on caring and generosity, not on price. Balance and reciprocity may play in a relationship, but are not the basis of it. One does not make deals for love. There are other words for that.

Back in the industrialized world, few of our market relationships run so deep, nor should they. By necessity much of our relating is shallow and temporary. We don't want to get personal with an ATM machine or even with real bank tellers. Friendly is nice, but in most business situations that's about as far as we want to go.

But relationship is a broad category: broad enough to contain all forms of relating—the shallow as well as the deep, the temporary as well as the enduring. In the business culture of the industrialized world, Sayo said, we barely understand relationship's full meaning or potential. And we should. Doing so would be good for business.

So he told me our next assignment was to unpack and study another thesis: *Markets are relationships*.

Containment Versus Relationship

IN FACT, "RELATIONSHIP" IS ALREADY THE MIDDLE NAME OF A multi-billion-dollar industry called customer relationship management (CRM).[3] The father of CRM is Thomas Siebel, whose book, TAKING CARE OF BUSINESS (2002), outlines "principles of eBusiness" that include "Know your customer," "Personalize the customer experience," "Optimize the value of every customer," "Focus on 100 percent customer satisfaction," and so on.[4]

The first problem here is that no customer can be 100 percent satisfied by services built for populations rather than for individuals. Companies (at least the ones that can afford CRM systems) interact with "the customer" and not with you. The common functions of CRM—marketing automation, sales force automation, call centers, lead generation, direct marketing, and other forms of management—are instruments of generalization. They treat customers as templates. CRM can improve those templates, developing better guesswork about what you might want; but it's still guesswork. It may be personalized, but it's not personal.

The second problem is that CRM would rather have captive customers than free ones. Every CRM system maintains what in the computer industry we call "silos" or "walled gardens." In Chapter 5 of CLUETRAIN ("The Hyperlinked Organization"), David Weinberger calls them "Fort Business." They are the containers companies build to "retain" customers they have "acquired," and where they "control," "own," and otherwise "lock in" those customers. Silos take the form of frequent flyer programs, publisher subscription lists, "memberships" of many kinds and—perhaps worst of all—"loyalty cards." Today nearly every grocery store, department store, bookstore chain, hotel chain, airline, and car rental company has its own "loyalty program," each a silo of its own. I have one friend who carries around a key ring the size of a necklace, strung with many dozens of little "key tags," each with its own corporate logo and personal barcode. All these systems are exclusive and incompatible with other loyalty programs. All are designed to increase "switching costs" and maintain other inconveniences for the customer.

The third problem is that CRM isn't built to deal with free and independent customers. Even on the Internet, where individuals enjoy far more independence and autonomy than ever, there persists an Industrial Age belief that "empowering" customers is an outside job—one only vendors can provide. There is little if any recognition that customers have any native power at all. All customer worth derives from cash, credit, and privileges granted by vendors. *The best customer is a captive one*, the assumption goes. Hence a "free market" is Your Choice Of Silo. We have been living with this belief for so long that we can hardly imagine any other market condition.

The novelist William Gibson once said, "The future is already here—it's just not evenly distributed." Customer service is a great example of an undistributed future. Companies like eBay and Amazon provide amazing and wonderful services. But try moving your eBay reputation to Amazon, or your movie reviews from Amazon to Yahoo. Or try "relating" to any large company in ways other than those the company provides. You can't because they control the whole thing. Your "relationship" is one they define, on their terms alone. Your choice is to live inside their silo or go find another one.

There is a good reason why CRM systems haven't made more progress: because they can't. Even if they all improved their relating ability a hundredfold, they'd each be doing it differently. They would still have their own ways of making you talk. It's like having to speak Swahili to one company, Greek to another, and Urdu to a third. We're stuck in the same kind of world that online companies occupied in 1990. The Internet existed then, but it was confined to government bodies, universities, and research institutions. Out in the commercial marketplace your only online choices were the likes of Compuserve, AOL, and Prodigy. Each was a silo with its own ways of organizing content and controlling communications. For example, each had its own email system, so you couldn't send messages from one silo to another. Your data wasn't even your own. It lived inside the system, where most of it got erased after a while anyway.

CRM systems are still stuck in this kind of world. It isn't a problem companies or their CRM suppliers can fix by themselves, any more than Compuserve or AOL could fix themselves before the Net came along. Only customers—that's us—can make CRM fully compliant with the Internet. Only we can prove that free customers are more valuable than captive ones.

VRM Meets CRM

AFTER GETTING MY *MARKETS ARE RELATIONSHIPS* ASSIGNMENT from Sayo, I began looking for technological developments that would do two things: (1) advance customer independence and (2) enable customers to engage vendors in ways that would be good for both sides. This drew me into conversations with developers in the digital identity movement. The first of these was Andre Durand, a prime mover behind Jabber instant messaging and its silo-spanning protocol, XMPP (both the brainchildren of Jeremie Miller, who has moved on to revolutionizing search, among other projects). Andre convinced me that individuals need to be in control of their own digital identity information, and that this control will be the first step toward full empowerment for customers, and the elimination of guesswork by vendors about what customers want.

By 2005, I had become a prime mover as well—in the "user-centric" digital identity movement. There are many development communities within that movement, all of which share the belief that individuals need to be at the centers of their own digital lives, and not peripheral dependents either of vendors or identity providers—which in many cases are the same thing. My work as a founder of the Identity Gang (yes, it was called that) brought me to the attention of John Clippinger, a senior fellow at Harvard's Berkman Center for Internet and Society. John offered Berkman as a clubhouse for the Gang (which has since become part of Identity Commons), and in the summer of 2006 I became a fellow at Berkman as well.

By this time I had begun to realize that identity control needed to be *user-driven* and not just "user-centric" (a distinction later clarified[5] for me by Adriana Cronin-Lukas)[6] I also realized that identity data was just one among many forms of data that needs to be under the individual's control. In the digital world we are still only beginning to build, customers need a whole box of tools that will give them new and useful ways to engage with vendors. With their new hammers, saws, nails, and wrenches, customers can work alongside vendors, breaking down silos and replacing them with a truly open marketplace.

The name for this toolbox is Vendor Relationship Management (VRM). A variety of development communities are already building tools for that box. Guiding them is ProjectVRM, which I run at the Berkman Center.

We didn't start by calling it VRM. That name came along in October 2006, during a Gillmor Gang podcast.[7] I was explaining the project and its goals to other gang members when one of them—Mike Vizard—clarified matters by calling the effort VRM. The rest of the gang ran with it, and the baby got named.

It's not a perfect name. "Vendor" is a business-to-business term ("seller" might work better in the retail world), and the scope of the challenge includes all relationships between individuals and organizations. For example, Britt Blaser and others are working on something they call GRM, for government relationship management. Some suggest drawing a circle around all forms of relating and calling the category RM, for relationship management. We'll see how that goes. Meanwhile, all VRM efforts share a few common purposes:

1. Provide tools for individuals to manage relationships with organizations. These tools are personal. That is, they belong to the individual in the sense that they are under the individual's control. They can also be social, in the sense that they can connect with others and support group formation and action. But they need to be personal first.

2. Make individuals the collection centers for their own data, so that transaction histories, health records, membership details, service contracts, and other forms of personal data aren't scattered throughout a forest of silos.

3. Give individuals the ability to share data selectively, without disclosing more personal information than the individual allows.

4. Give individuals the ability to control how their data is used by others, and for how long. This will include agreements requiring others to delete the individual's data when the relationship ends.

5. Give individuals the ability to assert their own terms of service, reducing or eliminating the need for organization-written terms of service that nobody reads and everybody has to "accept" anyway.

6. Give individuals the means for expressing demand in the open market, outside any organizational silo, without disclosing any unnecessary personal information.

7. Base relationship-managing tools on open standards, open APIs (application program interfaces), and open code. This will support a rising tide of activity that will lift an infinite variety of business boats, plus other social goods.

All these will also give rise to...

The Intention Economy

WE WROTE THE CLUETRAIN MANIFESTO WHILE THE DOT-COM BUBBLE was still gassing up. You might say CLUETRAIN was an appeal to sanity during a period of financial euphoria. It certainly felt that way at the time.

The dot-com bubble popped right after the book came out in January 2000, but that wasn't the only bubble that bothered us. We were also out to pop a much bigger bubble that had been growing since the dawn of commercial media. That bubble was advertising.

Unlike the dot-com bubble, the advertising bubble contains a large amount of real business. And advertising, like all bubbles, continues to grow at a suspiciously rapid rate. I'm writing this in January 2009, a time when financial news is bad across the board. This same month Google has revealed that its advertising sales revenues rose 18 percent in the last quarter, over the year before. According to PricewaterhouseCoopers, worldwide advertising spending will pass half a trillion dollars in 2010.[8]

Trust me: It's still a bubble. So is the rest of the "attention economy" that includes promotion, public relations, direct marketing, and other ways of pushing messages through media.

The attention economy will crash for three reasons. First, it has always been detached from the larger economy where actual goods and services are sold to actual customers. Second, it has always been inefficient and wasteful—flaws that could be rationalized only by the absence of anything better. Third, a better system will come along in which demand drives supply at least as well as supply drives demand. In other words, when the "intention economy" outperforms the attention economy.

Let's unpack all three of those reasons.

Number one: Customers and consumers in the attention economy are different populations. Customers are companies that buy advertising, PR, and other ways of targeting messages. Vendors are the media and agencies. You and I, as consumers of media and messages, pay nothing for advertising. We might subscribe to some newspapers and magazines, but in most cases we're not buying those for the advertising. (There are exceptions, such as fashion magazines, where ads are almost a form of editorial content; but in most cases advertising isn't what gives media their intrinsic value.) In fact, most advertising is something we tolerate more than demand. (As a journalist I can say the same about most PR, even though there are notable, even noble, exceptions.)

Thus the attention economy takes place almost entirely on the supply side of the marketplace. The demand side only becomes involved when individuals respond positively to an attention-grabbing message. But even when we do what an ad message wants, we are not paying for the message itself, meaning that we are not involved in the transactional part of the attention economy. This fact minimizes our influence over that economy and detaches us from it. We remain mere consumers: economic animals that Jerry Michalski calls "gullets with wallets and eyeballs."[9]

This detachment manifests as an operational split inside media organizations. That split is between what newspapers and magazines call editorial and publishing. In the online world that split is between content production and the advertising that supports it. In the old media world there was a "Chinese wall" between these two sides of the business. Reporters and editors did their best to work in isolation from influence by the mechanisms that paid for their work. In the online world the Chinese wall has cracked or disappeared. This is why and millions of website owners and bloggers (and sadly, many traditional media as well) now tailor their content for SEO (search engine optimization), so they can attract placements from Google or its competitors in the online advertising business.

As we pointed out ten years ago in this book, Alvin Toffler wrote about this split in THE THIRD WAVE (1980). He said we all had an "invisible wedge" in our heads, dividing our producer selves on one side from our consumer selves on the other. He said this wedge "ripped apart the underlying unity of society, creating a way of life filled with economic tension, social conflict, and psychological malaise." That wedge is still there. And the advertising bubble on the producer side throbs in our collective heads like a giant aneurysm.

Number two: The whole attention economy, especially advertising, is a low-percentage game. John Wanamaker famously said, "Half the money I spend on advertising is wasted; the trouble is I don't know which half."[10] But even this was delusional. Nearly all advertising falls on deaf ears or blind eyes. Yes, some small percentage of advertising messages get through and change minds or cause sales. But the proven success rates still involve odds in the lottery range.

The Internet changed this game by adding something advertising had lacked in the old analog world: accountability. As a digital medium the Net can add tracking and targeting data to advertising messages, and use that data to guide future messages far more carefully and accurately. Google especially deserves credit for making advertising accountable. Today, with Google's Adwords and Adsense, advertisers pay only for click-throughs. Eventually advertisers may pay only for actual sales. Google also deserves credit for moving the wasteful parts of advertising off of paper and airwaves and onto server farms. But even online advertising is still a guesswork game, and guesswork means waste—of time, attention, cycles, pixels, rods, and cones. And we're all still targets, not correspondents. There's little or no conversation (literal or metaphorical) between seller and buyer. So, even if we make advertising better and better, it's still guesswork. Improving the odds doesn't change the fact that advertising remains a game of chance. Nor does it eliminate irritation to the customer. Reducing a pain in the ass doesn't make it a kiss.

Number three: Something better will come along. Namely, the intention economy. This is the economy that will grow around what customers know they want and are ready to pay for. There is plenty of this in the world already. What it needs is a way to reach vendors who have no way to listen and instead are grasping at wallets with advertising. Creating the intention economy is one of the jobs we've taken on with VRM, in particular with a new market ritual we call the personal RFP. The term RFP (request for proposal) is a standard protocol by which buyers reach sellers in the B2B (business to business) world. RFPs have been absent in the B2C (business to consumer) world, for the simple reason that the latter goes only in one direction. The personal RFP will go in the other direction, so customers (no longer mere consumers) can tell the marketplace what they want and get back helpful responses, one of which will result in a sale.

For example, you might say, "I need a stroller for twins in Ann Arbor in the next four hours," without having to surf from site to site looking for sellers who have the product you want, and without having to do all your shopping inside some vendor's silo. Here *real* loyalty relationships can be helpful. For example, your RFP might go first to all the vendors with which you have relationships before it addresses the rest of the marketplace. It might also go to new classes of intermediaries, which will emerge because helpful services always show up where there is money ready to be spent.

The world today is full of MLOTT (money left on the table). Right now, for example, I'm ready to spend hundreds of dollars on an LED display with a resolution of 1920 x 1200 or better that will work with both a Lenovo laptop and a MacBook Pro, both of which have DVI connections. I've done a bunch of looking on search sites and I've posted a query on Twitter. Sometimes this works, but not this time. I'm still fishing. I want something better. I want to advertise my intention and have the business come to me. And I don't want to do it inside some website. I want to do it in the open marketplace. And someday I will be able to do that. I know, because VRM developers are working on it.

The attention economy will not go away. There will still be a need for vendors to promote their offerings. But that promotion will have a new context: the ability of customers to communicate what they need and want—and to maintain or terminate relationships. Thus the R in CRM will cease to be a euphemism. This will happen when we have standard protocols for all three forms of market activity: transaction, conversation, and relationship.

Transaction we already have. Conversation we are only beginning to develop. (Email, text messaging, and other standard and open protocols help here, but they are still just early steps.) Relationship is the wild frontier. Closed "social" environments like MySpace and Facebook are good places to experiment with some of what we'll need, but as of today they're still silos. Think of them as AOL 2.0. (And that's not a knock on AOL, which was a pioneer in its day, too.)

When we finish building the intention economy, the Mona Shaws of the world can approach whole markets with tools of demand that engage and improve the mechanisms of supply. When that happens, the heartless customer service desk will be safe from Mona's hammer—because both will be gone.

Notes

1. http://www.washingtonpost.com/ wp-dyn/content/article/2007/10/17/AR2007101702359.html.

2. http://www.insidenova.com/isn/news/local/article/taking_a_hammer_to _comcast/2772/.

3. Gartner Says Worldwide Customer Relationship Management Market Grew 23 Percent in 2007: http://www.gartner.com/it/page.jsp?id=715308. The story also says CRM was already an $8.1 billion business.

4. A source for the quotes: http://www.itstime.com/aug2001.htm. The Amazon entry: http://www.amazon.com/Taking-Care-eBusiness-Thomas-Siebel/dp/3478248809/ ref=sr_1_1?ie=UTF8&s=books&qid=1229382233&sr=1–1.

5. http://blogs.law.harvard.edu/vrm/2008/04/28/vrm-is-user-driven/.

6. http://www.mediainfluencer.net/2008/04/two-tales-of-user-centricities/.

7. History repeated itself here. On December 31, 2004, the Gillmor Gang podcast brought together the Identity Gang, which grew to become the Internet Identity Workshop.

8. http://www.marketingvox.com/pwc_entertainment_and_media_to_reach _18_trillion_advertising_521_billion_in_2010–022025/. The sources here are behind PwC's paywall.

9. In Chapter 4 ("Markets Are Conversations") of *The Cluetrain Manifesto*, we wrote, in the words of industry analyst Jerry Michalski, a consumer is no more than "a gullet whose only purpose in life is to gulp products and crap cash." Jerry has since said he prefers "gullets with wallets and eyeballs."

10. http://www.quotationspage.com/quote/1992.html.

But How Does It Taste?

I did toy with the idea of doing a cook-book. The recipes were to be the routine ones: how to make dry toast, instant coffee, hearts of lettuce and brownies. But as an added attraction, at no extra charge, my idea was to put a fried egg on the cover. I think a lot of people who hate literature but love fried eggs would buy it if the price was right.

Groucho Marx

YOU HAVE AN OVEN? MOVE ONE OF THE RACKS TO THE MIDDLE position, set the heat for 350°F (gas 4 or about 175°C for you folks on the right side of the pond), and turn it on. Take a half pound of unsalted butter and the same amount of a good dark chocolate, a chocolate you like to eat. Melt the butter and chocolate together in a bowl over some gently simmering water, stirring the mixture a bit. Crack four large eggs into another bowl; add a half teaspoon of salt, two teaspoons of a nice vanilla extract, and a cup each of dark brown and granulated sugar. Stir the sugar and egg mixture together, and when the chocolate and butter are melted, stir them in too. Fold in a cup of all-purpose flour.[1] Find a 13 x 9–inch metal pan, press a piece of aluminum foil into it, one large enough to hang over the ends, and grease it with a leftover wrapper from the butter. Pour the batter into the pan, put the pan in the oven, and bake for forty-five minutes. Scrape the bowl and lick the spatula.

When your oven timer dings, take the pan out and let it cool on a rack. If you're really obsessive, you can let it cool overnight, covered, in the fridge, but I never seem to muster enough obsession to let it sit that long. Using the foil ends as handles, gently lift the results onto a cutting board, trim off the edges to nibble later, and cut the remainder into twenty-four brownies. Find a friend or two, a neighbor, your spouse or kids, sit them down in the kitchen and share some brownies. Pour some milk. Enjoy the conversation. A lot has changed in the almost ten years since we published CLUETRAIN, but there's still a satisfaction I get from baking, from cooking, from talking with people over food. I've come to realize I'm a tactile person. As much as I've used the net, worked in high tech and depended on technology for a living, I still need to make things with my hands, and still need to hear others' voices. Those needs are the drivers for my current gig, running a small, albeit determinedly clueful, chocolate business. (Yes, chocolate. See www.sethellischocolatier.com for the longer story.) I find irony in seeing my own career tending toward creating something perceived as completely bereft of high tech while the integration of the net into the fabric of my daily life continues to increase at a headlong pace. More on that in a bit, but first, to bring the book up to CLUETRAIN 2.0, let's look at some changes.

What's Different?

THE BIGGEST CHANGES WE'VE SEEN SINCE OUR FIRST PASS AT CLUETRAIN aren't really in technology. Yes, if you believe the pundits, Web 2.0 is the biggest thing since the Internet crawled out from under its rock, but I'm not as sanguine as some about the usefulness of the Web 2.0 label. We technophiles love labels, as they make it faster to classify and understand what's going on around us. Labeling also lets us dismiss things more easily, as I can discard entire categories in my personal taxonomy when they fall out of favor. The problem with labels is they cut off discourse and truncate the process of understanding and exploration. My answer to a toddler's "What's that?" query shouldn't be "It's a bird." It's easy to demonstrate superior knowledge by giving names to things, but it's a lot better for kids to be guided with questions: "What's it doing?" "Why do you think it's doing that?" "Would you like to do that?" (At least it was better for my kids. They haven't stopped asking questions. Your mileage may vary.)

Back to Web 2.0, it's not obvious to me if the Web 2.0 label wasn't born out of a premature desire to categorize our social media technology, whether to demonstrate newness, to get some of it funded, or to prove we're observant folks and we know what's going on, preserving some standing in our technology priesthood. Handles like "Web 2.0" are mostly useful in hindsight, and often only when the subject being labeled is fairly far back in our rearview mirror. I don't think we're out of the divergent, explorative phase of the development of social media, let alone converging on best practice implementations, to use the label with any accuracy.[2]

Almost everything "new" happening today has seeds in technology existing ten years ago. The social changes we're seeing are even harder to pin down with labels, as behavioral shifts are often multiple steps removed in time and effect from underlying technology. They come as slow infusions rather than sudden changes in taste. To exacerbate the perils of using our crystal balls in labeling the next big thing, I am (and probably you are, if you're reading this) a card-carrying member of the lunatic fringe of early adopters of technology. We're obviously not the primary audience for the toys we build and use, at least not if we're honest about how we're going to make money selling the toys, and how they'll be useful to "real" people.

It's more practical to think of social media as the use of the mundane, commonplace technology around us in the pursuit of a goal transcending that technology: fostering conversation and connection between people. Madge the manicurist, speaking for generations of Palmolive users, provided the best guidance for would-be spotters of social media innovation: "You're soaking in it."[3]

All that said, there have been a few changes since CLUETRAIN. The general outline of change revolves around some basic infrastructure improvements, and the resulting social ripples riding along with them. On the short list of change topics are access, data scale, usability, and participation.

1. *Access*. Data pipes have gotten bigger and faster, and more people are connected to them; 75 percent of the North American population has access to the Internet,[4] and 55 percent of connected homes have broadband access.[5] In addition, we're less tethered, with 84 percent of our population having cell

phones, as opposed to only 34 percent in 2000. The U.S. population has also gone from sending 12 million text messages each month when we published the manifesto to more than 75 billion per month now.[6] When we wrote CLUETRAIN, we were very aware of the economic and class-based disparities in access to the net. While those disparities still exist (and I'll get to more on usability, literacy, and age-dependent access, below) the increase in access since 1999 is staggering. We were flogging conversation, hoping people would show up, plug in, and start talking. They did.

2. *Data scale*. More and more information is searchable, and the odds of finding useful answers to my questions on the net have increased dramatically. Even as we're bemoaning how the net may be changing our reading habits and reducing our ability to focus, anyone who can get to the net uses Google, depending on it for everything from quick access to driving directions to being a surrogate memory for phone numbers, names, and facts. Ten years ago, the data wasn't there. There's been a mind-boggling increase in the amount of information available to us on the web since CLUETRAIN. The public web has grown from around a billion searchable pages in 2000 to more than a trillion today, a thousand-fold increase. The odds of the "long tail" yielding information about your products and your company from your customers—one of my favorite CLUETRAIN themes—have increased substantially. (Cautionary note: To find data on the net, we depend on people having the foresight and economic resources to keep it on disk, attached to a computer with a net connection. What happens when it isn't feasible for them to continue to maintain their part of the cloud? It's a risk worth thinking about.)

3. *Usability*. On the usability front, we have both good and bad news. Bigger, faster pipes and faster computers are responsible for faster graphics and more responsive interfaces. Similarly, a trend toward smaller applications and reusable chunks of fast, optimized software for our browsers and phones means our interaction with computers has the potential to be faster and less painful. For those of us who are the most computer literate, this is great news. My pages load faster, I can watch video more easily, and the applications I depend on have moved from my desktop to my browser to my phone.

The flip side of the usability issue is the continuing lack of support for the less sophisticated in our audience, both seniors and those with only basic literacy skills. The tendency towards complicated, flashy, nonaccessible interfaces is exacerbated, in part, because both the designers and early adopters of much new net technology are young and literate. We're leaving an increasing percentage of our population out in the Web 2.0 cold. The economic aspects of the digital divide get somewhat better as computers get cheaper. The literacy and ageist issues don't. If we're serious about getting as many people as possible under our conversational tent, we need to get serious about designing inclusivity into the web.

4. *Participation*. In 1999, conversation online was in newsgroups and indexed email lists, and captured in web pages maintained by the techno-obsessive among us. Blogging was in its infancy, and a blog was most likely a diary-like page of a website hand-maintained by a very determined writer. Today, thanks to easy-to-use applications and services like Blogger, Wordpress, and TypePad, more than 100 million blogs have been started worldwide. (Treat that statistic gingerly, as it's hard to tell how many authors continue posting to their blog once the novelty has worn off.) Conservatively, there are more than 75,000 blogs with sizable public followings, according to blog search engine Technorati. Google's blog search engine returns results from more than 2 billion blog postings. Considering the number of blogs allowing and encouraging comments by readers, that's a lot of CLUETRAIN fodder.

Blogs are one element of the conversational landscape we foresaw when we wrote CLUETRAIN, but social networking on the scale we see today was just a hazy premonition. In hindsight, one of the keys to accelerating the social changes we described in CLUETRAIN has been reducing or eliminating barriers to entry, reducing the friction stopping people from participating in conversations. Many social network apps will, with my permission, harvest contacts from my address book, pre-populating my list of online friends and also letting them know I've arrived, substantially improving the odds of my having someone to talk to. Twitter, notable for legitimizing tiny posts, "tweets" of 140 characters or less, as net conversation, prompts me to send a tweet each day, dramatically increasing the likelihood I'll participate at least once a day in the network. In addition, many sites use the side effects of my participation to build their shared data space and make it more useful; my purchases on Amazon add to stats showing what people have bought; Flickr feeds show me pictures others are uploading; Delicious, Digg, StumbleUpon, and other social bookmarking sites allow me to search large collections of links highlighted by others, or just see those from my friends or from people with similar interests; Facebook and MySpace widgets let me know what my friends are browsing and listening to.

As I recite these examples, it's obvious to me we haven't seen the zenith of conversational tools on the net. The sense of fragmentation in the social media space, of too many options and not enough integration and sharing, reinforces my sense we're still early in the innovation cycle. There are walls between my Linked-In universe, my Facebook universe, and my phone. Despite attempts to connect my dots of speech, I still perform the same tasks at each service site, gather and regather my set of conversational partners, identify myself over and over again with password after password. We need to be more fanatical in our elimination of conversational friction. I have some hope our understanding of the ground rules for computer-to-computer or service-to-service cooperation is getting sophisticated enough to allow us to begin to piece together the bones of the future "semantic web," the shared utility underpinnings needed to allow our tools to "do what I mean," or "do what I need."[7] But it ain't here yet.

Looking back at our CLUETRAIN track record, we've seen our prognostications of networked conversation vindicated. While we were optimistic about how quickly the added conversational volume would change the marketing and business landscape (and I leave calibrating that optimism to my fellow authors), there are certainly more people talking, and some may even be listening.

Bullshit Detection

I'M FASCINATED BY HOW WE DISTINGUISH "GOOD" SPEECH FROM noise. I still get the occasional email from a friend warning me the sky is falling because of a computer virus and exhorting me to take precautions. While a quick Google search with some text from the message and the word "scam" can often dispel their fears, I wonder how we train ourselves to trust and not trust, given the paucity of clues about an author's intent and bona fides provided in an email or on a web posting. On one end of the scale are Nigerian 419 and Irish sweepstakes scams. Nobody in my acquaintance still falls for them, I guess because they're blatant enough to trigger skepticism in even the most hopeful scoring-something-for-nothing reader.

When I'm shopping for a book, I read reviews and try to discern which of the reviewers are most like me; if I'm considering buying a technical book on chocolate or cooking, do they simply like the recipes or do they evaluate how well they work? Do they state their own background, and do so in an open, believable way, or are they self-aggrandizing and therefore seemingly (to me) less trustable? On Amazon, I put more trust in real-name reviewers, in the (perhaps misguided) assumption that those willing to write opinions under their own name will be constrained to be more honest. (Anonymity may help people to speak truth, but it doesn't help me to trust them.) Given a large enough quantity of prose from one author, whether named or under a pseudonym, we have an easier time calibrating their speech and knowing how to react to what we read from them.

As we went through the last presidential campaign and I read through the broadsides about one candidate or the other, I found myself asking, "Can this be real?" and googled for more info to help judge what I was reading. For instance, a letter from Anne Kilkenny, a voter registrar in Wasilla, Alaska, was published on the net, presenting a critical assessment of Sarah Palin's record from the perspective of someone who knew her and had lived and worked alongside her.[8] A conservative blogger speculated we were seeing the birth of an urban legend, and there was no Anne Kilkenny in real life.[9] Later in the day, she recanted, describing her subsequent email conversation with Kilkenny and admitting, perhaps grudgingly, that Anne was real.[10] That open documentation of skepticism and discovery of contradicting truth made her appear more trustable in a way simple protesting speech would not, and gave me a better sense of how to read her other writing.

The essence of trust in something I read comes from being able to discern a basic honesty and a sense of openness in the prose. The meme "authenticity" circulates among marketing wonks when discussing how to create effective corporate blogs, but it's not the same thing. We can tell when a person is writing in his or her own voice. If you're debating whether your corporate speech is authentic or not, you're probably already lost. I'm rarely wrong in deciding if I'm hearing a real person or not. My problem is flipping a coin on whether I trust a stranger to tell the truth, and if they're telling their truth, will it be truth for me as well?

As the skeptics among you have pointed out over the last ten years, our markets-are-conversations meme has an interesting underbelly: 99.9 percent of all conversation isn't conversation I want to listen to. That means 99.9 percent of all markets may be crap too (given that each of us will reject a different 99.9 percent). But it shouldn't be a surprise, as the wired world reflects the real world. We're awfully selective when we're spending our own nickels. How many of the tens of thousands of products in your neighborhood Safeway, Kroger, ShopRite, or Publix do you actually buy? 99.9 percent may be a realistic number.

Maps, Territories, and Real Conversation

I'M IN A REAL-WORLD BUSINESS. MY PRODUCT—GORGEOUS, luscious-tasting organic chocolate truffles—isn't something the Internet can help you sample. You can read the reviews, order the candy, and do the electronic commerce dance. But when it comes down to the moment of chocolate truth, you have to be in the same room as the product. (After 30-plus years of living with a work product my kids can only see by peering into a computer display, I've realized I like to create things people can touch. Go figure. Have some chocolate.)

In the three years it's taken us to plan and set up the business, I've had my eyes reopened to how useful and how limiting the net is. On the useful side, the net makes it easy to find possibilities for a business. We search for potential sources of ingredients, for machinery, for manufacturers of various supplies and equipment. But what we receive from a Google search are only possibilities. Almost every purchase transaction is preceded by numerous phone calls, sometimes face-to-face meetings, and lots more legwork than a simple Google search will accomplish. The world reflected in Google is a shadow world, hinting of connections and complexities for our business, but it doesn't gain substance until we have the real conversations leading to an exchange of ideas and goods.

Even when it comes to simple documentation, there's an incredible amount of information about the chocolate industry untouched by the net. While the Google folks pride themselves on getting umpteen thousands of Civil War medical records online, I own several shelf-yards of reference books from the past eighty years having, at best, only vague electronic ghosts in Google's online indices. Sometimes the most I can do is find the name of a publisher or author, and then set an automated eBay search for them, hoping sometime in the next six months my trap will spring, and I can buy a dog-eared copy of the, until then, unnamed manual for my library.

As far as finding people goes, I've learned the net is most useful, again, for finding starting points, phone numbers to call; not for answers, but to beg suggestions about who else I should ask about my problems and questions. I've managed to build a wide-reaching network of people more experienced than I, from whom I learn, to whom I return with more hard problems, and who call me when their difficulties are ones I can help solve.

Even when we're searching for news and clippings to help understand our industry, to see our competitors' new products, or just to understand trends, we're often confronted with bald statements of fact, without the underlying logic and subtlety of reasoned judgment we need to understand what we're reading. Again, more phone calls, more discussion, more real-world conversation.

I hope more thoughtfulness than amazement comes across as I write this. I've had the chance to experience an interesting shift from netspace back to realspace, one taking me back to thoughts of working for my dad as a potter, with no electronics, no wires, and no flashing lights, but plenty of conversation and physical trial and error. I hadn't realized how much I missed it, and it gives me an interesting foil for the frequent moments when the net still intrudes or our complex IT systems need debugging or a reboot is finally required.

It Still Doesn't Scale

19 Companies can now communicate with their markets directly. If they blow it, it could be their last chance.

22 Getting a sense of humor does not mean putting some jokes on the corporate web site. Rather, it requires big values, a little humility, straight talk, and a genuine point of view.

23 Companies attempting to "position" themselves need to *take* a position. Optimally, it should relate to something their market actually cares about.

25 Companies need to come down from their Ivory Towers and talk to the people with whom they hope to create relationships.

If you suspend a bit of disbelief and skip past the arrogant tone, there's a seductive simplicity to CLUETRAIN. We never said it in so many words, but we strongly implied that conversation is an easy thing. Open your mouth, say something engaging in your own, honest voice. If you had something interesting to say, we'd listen, join the conversation, and judge your company to be one of the clueful ones.

The details we neglected to address were manifest. I can quickly reel off a few, and I'm sure you can contribute your own:

- How can you change a large corporate structure, one safely "protected" by an active PR operation, into a nimble, conversational entity with thousands of clueful ambassadors?

- What happens to legal liability when one of your ambassadors tells a truth with repercussions?

- Where do we find the time to listen to all the shouting and answer every yawp?

- ...and so on.

Tackling the problem of conversation across the Internet/intranet membrane was always the one I envisioned when we talked about how to implement a CLUETRAIN communication strategy. My current gig pushes home that point, as I have to worry about how we get the word out about our business, how we

keep a customer's interest after he or she tastes our product, and how we maxi-
mize the potential for word-of-mouth connection to the next customer.[11]
However, when I look for companies succeeding at this, most of the ones I
come up with are essentially small businesses, with simple organizational underpinnings and a scale made manageable by an intrinsic limit on the number of
company participants in the conversations. Take a peek at South African wine
company Stormhoek and the work they've done with Hugh MacLeod, one of my
favorite food marketing examples.[12] They're engaging, are using a bunch of
media tools effectively, and have an unmistakable, comfortable style and voice.

Since I'm focused on small businesses almost exclusively these days, as I
started poking at the scalability of CLUETRAIN I asked David, Chris, and Doc
for examples of larger enterprises showing signs of tackling the issues around
conversational scale. The examples they suggested are, at their core, all demonstrations of how conversation might be made to work on the net.
Some samples:

- Amazon.com's ubiquitous reviews, where even our almost ten-
  year-old manuscript is still getting fresh comments. The site does
  a good job to help me become a smarter consumer of my current
  commodity of interest in a matter of minutes, letting me read
  about others' experiences with the product I want, and showing
  me what other people (presumably ones just like me) bought after
  browsing the comments.

- The Open Salon social media site, where they've created a site for
  people to post longish form stories, in the midst of a conversational community.[13] Their "Tippem" tip jar concept is interesting,
  but the jury's still out on it.

- Various hybrid journalism/blog sites, including HUFFINGTON
  POST, SLATE, and others, where often the volume of reader-
  authored comments on a story will far exceed the length of the
  story itself.

- Twitter and tweet aggregators, like election.twitter.com and apps
  like Tweetdeck, bundling, and organizing Twitter's text snippets
  into themed or categorized streams.

Looking at these, I'm struck by a basic similarity: They're all harnessing reader-generated content and reader conversations.[14] Very little of the conversation, if any, is between the people running the enterprise and their customers. They're all providing a venue for speech and usually something to talk about, whether products, news, prose, or gossip, but the enabling company is rarely in the midst of the conversations. There may be a small number of people from the enterprise participating in the conversations (Open Salon, Slate, etc.) or none at all (Amazon, at least not noticeably). The insiders don't have to carry the conversations; the outsiders, you and I, do that for them.

I'm not sure any of this gives me a silver bullet answer to questions about making CLUETRAIN scale. There are some hints here. Markets with multisided conversations, letting me hear multiple parties, not just one, are more interesting. Having a small number of people doing the talking reduces risk and makes the task of supporting conversation more manageable. Using speech generated by a community makes your efforts more interesting to more people, with less cost to you. But I'm still unimpressed with our protestations of scalability. Maybe only a small enterprise, and a small community, can really do the CLUETRAIN trick.

I'm reminded of why I liked working at Sun Microsystems in the 1980s and 1990s. The company, for better or for worse, was a sometimes loosely coupled collection of entrepreneurial groups, each with its own voice and customer connections. This worked when we talked to each other or had a strong manager guiding internal coordination, or when an individual group was under the "scale threshold" for conversational success. Only then could we successfully navigate connecting to the world on the other side of the conversational membrane. We were a bunch of smaller companies inside a big company.

Perhaps that's one of our answers to scaling. Keep it small. Conversations happen between individuals and among small groups of interested people. When the group gets too big, we break into smaller conversational circles. Collecting lots of speech gives us a matrix in which interesting conversations might happen, but without the round trip, without individuals listening to and responding to others, there isn't conversation.

I get very little back when sixty people at a time taste our chocolate. I've tried. I lose most of my opportunity to connect with customers because I can only talk to a few at a time. In my business, I need to watch people's reactions, person by person, and listen.

How does it taste?

Notes

1. I have a killer gluten-free variant of this recipe. Email me.

2. When asked for Google's take on Web 3.0 in mid-2007, Eric Schmidt replied, "Web 2.0 is a marketing term, and I think you've just invented Web 3.0." www.youtube.com/watch?v=T0QJmmdw3b0 (apologies to John Markoff and Nova Spivak!).

3. Hat tip to Jeff Cobb: www.jeffthomascobb.com/blog/2008/06/social-media-soaking-in-it/. See an early Palmolive video here: www.youtube.com/watch?v=_bEkq7JCbik. Jan Miner, a wonderful radio and stage star, played Madge in more than twenty-five years of Palmolive commercials. For those who don't know her from any other role, listen here: www.goldenage-wtic.org/05-JAN_MINER.MP3. She had a lot more range than just Madge!

4. Pew Internet & American Life Project: www.pewinternet.org/trends.asp. Many thanks to Susannah Fox at Pew, who answered her phone when I called five minutes from quitting time on the day before Thanksgiving and spent too much time answering my questions and giving me a tour of their excellent data. For data nerds only, see the Usage Over Time spreadsheet at that same link. It's quite nice.

5. Internetworldstats.com and Nielsen Online.

6. June 2008, CTIA.

7. For the brave, try this overview by Tim Berners-Lee from 1998. We've been chewing on this for while: www.w3.org/DesignIssues/Semantic.html.

8. www.snopes.com/politics/info/kilkenny.asp.

9. Cindy Kilkenny, no relation to Anne, http://fairlyconservative.com.

10. http://fairlyconservative.com/the-race-for-president/a-chat-with-anne-kilkenny-from-alaska/.

11. Turns out that trial, getting people to taste the chocolate, is the best way to drive sales. Not quite as scalable as I'd like, but effective. It's so-o-o much easier to sell chocolate

than life insurance or annuities or software. No conversations about ROI, cost of ownership, or maintenance. Just "Here, taste this." And then stop talking.

12. Stormhoek's site is www.stormhoek.com/blog/. Hugh can be found at www.gap-ingvoid.com.

Some social media marketing advice from Chris Rawlinson and the folks at Stormhoek on the occasion of one of their neighbors launching on the web:

1. Doing a group blog is difficult, a lot harder than doing a single-author blog. So, each author needs to find his voice and role within the blog. Make your voice authentic.

2. Don't just pimp your stuff.

3. Decide what conversation you want to own.

4. Short posts delight readers.

5. Nobody cares as much as you do.

6. Do a video.

7. Link to people you like and even the ones you don't like, if they are doing something worthwhile.

13. http://open.salon.com/.

14. If we're honest, a lot of this isn't conversation but one-sided speech. For instance, I read Amazon reviews, I act on what I read, but I don't speak back. I'm not quite comfortable calling this kind of captured opinion "conversation." Valuable, undeniably; written by real people, yes, but not quite conversation. It takes a round trip or two. Amazon's review speech is very different from the under-the-corporate-radar conversations I might use to find a car dealer I can trust, learn how to fix something, or help my daughter pick a college.

Obedient Poodles for God and Country

CHRIS LOCKE

*If I can't hold on to you, leave me
somethin' I can hold onto.*

Jonny Lang, "Lie to Me"

IT WAS 1966 AND I WAS AT THE MONROE GRILL IN ROCHESTER,
New York. Headlining that night was Buddy Guy, the monster Chicago bluesman,
along with his sideman, Luther "Snake" Johnson. After the last set, Luther and I
sat up till damn near dawn talking, swapping lies, and drinking cheap whiskey. I
think it was called the Monroe Grill. I was already pretty wasted when I got
there, and it looks like it's too far back even for Google. Lost now in the shifting
mists of time. Point is, the band was cooking that night and the place was rock-
ing, everybody in the groove. It was insane.

Buddy had about two hundred feet of cord attached to his guitar, which
seemed a bit excessive—that is, until he jumped off the stage in the middle of a
number, threaded his way through the crowd—and left the building. It was snow-
ing outside that night. Cold as hell. People walking down the street were seeing
this crazy bastard wailing on an electric guitar that wasn't making any sound at
all. Probably one of your psychotic street people. Shaking their heads. Sad, really.

But inside the club, the blues kept coming big-time thanks to that long umbilical cord. He was totally blowing us away. And when he finally came back inside, still jamming, never missing a beat, people were yelling, cheering, jumping off tables, flipping out.

That's my metaphor for the Internet ten years ago when we were just starting to talk about what would become THE CLUETRAIN MANIFESTO. Business rushing by on its way to the next interminable meeting, looking askance at some crazy street act, while we—"audience to each other," as I wrote back then—were safe and warm inside a very different, if virtual world, having the time of our lives.

At least it seemed safe. Like a safe bet, anyway, that things were really going to change this time. How could they *not*, with all these hooked-up barbarians straining at the gates?

Then the "Internet bubble" burst. Well hey, no surprise there. Hadn't we been saying that everything business was doing on the net was nuts? And then, a year and a half after CLUETRAIN appeared as a book, those planes plowed into the World Trade Center. As it transpired, the world was not so flat as some had dreamed in the opium dens of gung-ho globalization. Whoops. It seemed we'd forgotten there was a real world out there, beyond the insular navel gazing of Silicon Valley and the Washington Beltway.

Between the Bubble and the Towers, business cooled considerably in its fanatical whoring after e-commerce billions. All of a sudden, the net was anathema. Pretty much nobody outside the net itself wanted to write about it. Business didn't want to hear about it. I wrote a book called GONZO MARKETING: WINNING THROUGH WORST PRACTICES that came out a month after 9/11. But nobody wanted to read much of anything in those dark days that wasn't about anthrax or data security. I was bummed. To quote from REPO MAN: I blame society. *J'accuse!*

> Duke: **The lights are growing dim, Otto. I know a life of crime has led me to this sorry fate, and yet I blame society. Society made me what I am.**
>
> Otto: **That's bullshit. You're a white suburban punk just like me.**

But the net clearly hadn't disappeared or gone on vacation. Thanks to the Internet Movie Database and YouTube (which did not yet exist back then), those immortal lines are preserved for later generations—to puzzle over, if nothing else.

Eventually the money came back as it always does, having little clue where else to go. There was a new wave of venture cash for Internet startups, but this time less flamboyant, more circumspect in their approach to those market eyeballs and their metonymously groaning wallets. This time there was a greater focus on what Wall Street wanted to hear more of: that magic mantra, *monetization*. At the same time, with no apparent cognitive dissonance, there were those companies who outwardly genuflected in the vague direction of CLUETRAIN. "We get it!" they wanted us to know. Uh-huh. Sure you do.

And of course there was Web 2.0 and social networking and all manner of signs and wonders in the network firmament. Some of this stuff was real. Some of it even important. I guess. But I was losing interest. Which is not to fault the developers who've made all the cool tools and toys we see on the web today. Why, the buzz from Twitter alone could power Milwaukee for a solid year!

My loss of interest was a personal thing. Not only was there a world *out there*, there was also—who knew?—a world *in here*. The Towers became a tarot card metaphor for a similarly catastrophic sequence of events in my life back then. Which, don't worry, I won't go into here. Suffice it to say, my Towers in rubble, I looked to other horizons.

And is it any wonder? Marshall McLuhan once said that each new communications medium is always perceived—mistakenly—in terms of its predecessor: photography as a new form of painting; film as a new form of stage theater; the Internet as a new form of TV. Too right. And yet...once the New Thing is seen as *genuinely* new, how long can we marvel at its newness? How long can we enthuse over its many splendors?

When telephones were a new thing, people would comment on the device itself. "Why, I can hear you as if you were in the next room!" Or, "It's for you. Hurry, it's *long distance*!" When television was new, someone might have said, "Those guys on DRAGNET sure don't look like they did on the radio!" For everything that is gained, something is lost.

When the interstate highway system was new, back in the days of Eisenhower, I suppose there were lots of editorials in local newspapers about how the downtown business district was being affected. But it would be quite some time before scholars started writing about the impact of the interstate highways on the *culture* of the United States. How, for instance, it fueled the proliferation of the Beat generation kicked off by Kerouac in ON THE ROAD, or more prosaically perhaps, the mission from God taken up by THE BLUES BROTHERS:

> Elwood: **It's 106 miles to Chicago, we got a full tank of gas, half a pack of cigarettes, it's dark, and we're wearing sunglasses.**
>
> Jake: **Hit it.**

Notice that Belushi did *not* say, "Wow look, they're pointing that camera at us and, thanks to this brilliant film technology, millions will see us acting like fools!"

Yet how many Internet conversations—of which markets purportedly consist—are about the very technology enabling them? When it isn't about cats wanting cheeseburgers, that is. Come to think of it, the "I can has cheezburger?" meme (you could Google it) makes a point: Some of the funniest, wackiest, most brilliantly creative stuff on the net is—and was always—remarkable and refreshing precisely because it's *not* about the carrier medium. No, it's about jumping off the tables while Buddy's playing outside in the snow and the quotidian, pedestrian world passing him by out there still doesn't get why that's the most wonderful thing *ever*! Or at least tonight.

Notice that the audience did *not* say: "Behold! The electricity from his amplifier traveled all that way without resulting in any loss in decibels!" No, the audience said, "Let's boogie!"

But to a large extent, we are still talking on the web about pipes and wires, about online business schemes and technical methodologies, about the tools themselves. Not that there's anything wrong with tools. You discover fire, you make a bow, you chop obsidian to make an arrowhead. Sooner or later, you have to *do something* with those tools. Shoot something. Eat it. Just ask Sarah Palin.

Which brings me to another class of things in the larger world beyond the introverted Internet: politics. Since 9/11, many of us have grown increasingly aware that the vaunted "Internet revolution" has largely served to distract from the erosion of our liberties, the corruption of our laws and social covenants, the theft of our financial and environmental wealth, and the steady rise of a larcenous cadre of unprincipled political thugs. Not that I have strong feelings in the matter.

I am writing this mere days after the election of Barack Obama. The net had a lot to do with Obama's win. It was crucial in getting out the vote, vital to the unprecedented level of fund-raising. YouTube did not exist during the 2004 election campaign but played a huge role in 2008. And the diehard pals of John and Cindy, of Mooseburger Mom and her First Dude, just didn't seem to have the juice, the chops, the manic *intensity* of Obama's shock troops—among whom I counted myself.

In 2000, when CLUETRAIN was first published, I wasn't so political. I did get to meet Al Gore. I asked him if he'd ever read the First Amendment. A roomful of editors at the NEW REPUBLIC offices in Washington, D.C., took my cheekiness as an opportunity to check their manicures, no one venturing Word One. Gore had been explaining to me the upside of the Communications Decency Act, which would have censored the Internet. I'll give you an inconvenient truth: Tipper. And about a billion "explicit lyrics" labels. I'm with Eminem on that one, and what he said in "White America": "Fuck you, Tipper Gore. Fuck you with the freest of speech this Divided States of Embarrassment will allow me to have." But he ended that song: "I'm just playin', America. You know I love you." Yeah, me too. So I held my nose in 2000 and voted for Al. What a rush, eight years later, to be able to do more than "flush a protesting toilet and run," as Lawrence Ferlinghetti once wrote with reference to the guy who signed—well, would you look at that!—the National Interstate and Defense Highways Act of 1956.[1] (You thought I was just rambling, didn't you? You freaking beatnik. But see: We're still on the road!)

On the night of November 4, watching the throng in Grant Park waiting for Obama to appear, I was at first confused. Why were the CNN cameras focusing only on the crowd? Why not turn them toward the podium? Clearly something was about to happen—all these people were cheering wildly! Then it dawned on me. They were watching the huge monitors that had been set up for the event. America was cheering *itself*. What a beautiful moment.

Another difference from 2004: All this was on the net. I haven't had TV for eight years, but I was able to watch—in real-time video on my laptop screen—the conventions, the debates, and the victory celebration in Chicago. Impressive, magnificent technology. The tools have gotten better, the bows bigger and stronger, the arrows so much sharper. Now to kill something! Metaphorically speaking, of course.

As I watched that crowd in Grant Park, it seemed Oprah was getting a *lot* of face time with the camera. Simply put, Oprah spooks me. To explain why, I need to unpack what I did with those tools while my personal Twin Towers were still smoking wreckage, awaiting the psychic garbage barges to haul them off to China before anyone could forensically examine what remained of my allegorical WTC 7. Less simply put, Oprah strikes me as the outward and visible velvet-gloved avatar of some ethereal and ghostly iron fist.

Recall our interstate highways trope, and reflect that something like the delayed cultural impact of that is happening today. Back then, we could look at eight-lane highways and imposing cloverleafs, and understand very little of how they would come to change America. It would take some sort of paranormal prescience to grasp the future cultural consequences of such large develop-ments as interstate highways and "information highways" as they are taking place. Today we can look at the state of the web and see only faint outlines of the culture this technology is enabling.

And it's impossible to ask the interstate highway question about America without looking long and hard at religion. Even the money says "In God We Trust." When it goes *that* far, you know you're dealing with a deeply "faith-based initiative."[2] The phrase is from some 2001 "remarks" by George W. Bush, but yes, Virginia, there were Evangelicals going way back—even before the Internet!

Back around 2004, I started poking around in the backwaters of the nine-teenth century. I was trying to answer a personal question: How had Boulder,

Colorado, where I live, come to resemble a medieval mummer's show? The
place is saturated with the irrational. As long as it's magical, mystical, or mysterious, you'll find it here: astrology, Reiki, tarot, gnosticism, shamanism, Wiccan
goddess worship, Castaneda-inspired Tensegrity, Jungian "ally work," Lucid
Dreaming, Eckankar, integral "philosophy," Promise Keepers, the Jack Kerouac
School of Disembodied Poetics, harmonic convergence, Shambhala training...
the list goes on and on. The faux-quantum-physics movie WHAT THE BLEEP DO
WE KNOW? got its start here with funding from local software entrepreneur
William Arntz, whose guru channels a 35,000-year-old Lemurian warrior named
Ramtha.[3] Jello Biafra, lead singer for the Dead Kennedys, was originally from
Boulder. He once said:

> Boulder was a magnet for all kinds of so-called movements and
> escapist snake oil salesmen that would now fall under the umbrella
> of what they call New Age. And I thought, oh my God, this sort of
> apathy and putting your brain to sleep and becoming an obedient
> little poodle is a one-way ticket to fascism. It's how people allow
> themselves to be taken advantage of by aspiring dictators.[4]

Is Boulder unique in this regard? Hardly. Many suspect—I did at first—that
this magical mystery tour began in the 1960s with psychedelic drugs and the
Beatles visiting the Maharishi. Not even close, as it turns out. Before that there
was all manner of spooky Jungiana, including an anti-Semitism that Jung's fan
club has repeatedly tried to sweep under the rug, along with the great man's
words of praise for Hitler, whom he called "the mouthpiece of the gods"[5] and a
"truly mystic medicine man."[6] (The links are to TIME magazine reports of 1936
and 1939, respectively.) That worked right in with Heinrich Himmler's fascination with the occult, and his SS-sponsored exploration of Tibet, which, going
back further still, was a belated search for what Madame Blavatsky described,
in her SECRET DOCTRINE of 1888, as the "Aryan root race."

But the fascination with otherworldly weirdness goes back further than
that. For instance, to the "burned-over district" of upstate New York, so-called
not because of forest fires, but as a result of the Second Great Awakening, the
revivalist religious movement that gave rise to so many cults there. The most
famous of these was Mormonism, an American upstart religion that saw Native
Americans as the spawn of a wicked seed, and refused to ordain black priests

until 1978.[7] From the same fertile region in mid-nineteenth century came the indigenous American hoax of Spiritualism—talking to the dead by table rapping in Morse code. And earlier still, in Maine in 1838, a watchmaker named Phineas Quimby listened to a lecture by a disciple of Franz Anton Mesmer and started a quasi-religious movement that has come to be variously called Mental Healing, Mental Science, Mind Science, and New Thought—hard to distinguish from what we would today call New Age. America has been awash in such stuff for a very long time.

Remember, I was asking interstate highway–type questions about how American culture—and by extension and export, the world's—had *already* changed because of something that started happening over a century ago. Remember too that in the age of e-commerce, export entails the Internet. To document my ruminative speculations in the course of these researches, I started a blog I called Mystic Bourgeoisie.[8] At the top of the main page it says, "The unlikely story of how America slipped the surly bonds of earth & came to believe in signs & portents that would make the Middle Ages blush." My hypothesizing begins roughly with Emanuel Swedenborg (1688–1772) and Ralph Waldo Emerson (1803–1882). Note that this is about as far from CLUETRAIN's focus on business and marketing as it's possible to get.

Which is not to say that Mystic Bourgeoisie is any stranger to the tools and conversations CLUETRAIN championed. Without the incredible research tools provided by Google Books and Amazon (perhaps especially the collective knowl-edge and whacked-out opinions aggregated in millions of reviews there), with-out Google search more generally, without Blogger for publishing, without email for distribution (i.e., hassling my poor subscribers to actually *read* what I publish) and extended behind the scenes exchanges of tips (I get lots of them) and thoughtful feedback...in other words without the *web*, this project would simply not be viable. The web allows independent scholars like myself to accomplish what was once possible only for tenured professors in large univer-sities with herds of fawning grad students at their command (and control). Plus—and it's a big plus—I don't have to be particularly politic or polite.

But hold on. Mystic B is not entirely divorced from the first edition of this book. A major focus of CLUETRAIN was the issue of authority. As in command-

and-control hierarchies. As in authoritarian management. Now the thing about authority is that it always wants to capitalize itself. I don't mean with venture capital. I mean like this: Authority. And it tries to achieve this by referring its "natural" mandate to an increasingly Higher Power. Pretty soon, presto: GOD!

Swedenborg and Emerson were looking for the highest of all possible Higher Powers. The former was quite possibly mad—the jury is still out, though he does seem to have awakened the neighbors on more than one occasion by loudly remonstrating with various visiting demons. Emerson and his Trance-and-Dental pals decided to look in a less Christian direction: the Mystic East. Much of what we think we know today about oriental religions comes from frontier Protestant misreadings of bad translations of the Hindu classics. Emerson's pantheist NATURE was the velvet glove for his notions of rugged individualism and "Self-Reliance"—the iron fist of Manifest Destiny.

Which brings us back to Oprah in Obama's front row. And why that scares me. Oprah has been the vector for many of those "signs & portents" Mystic Bourgeoisie alludes to. One recent example is her "Soul Series," which began with Oprah "co-teaching," via the web, with Eckhart Tolle, whose book, A NEW EARTH: AWAKENING TO YOUR LIFE'S PURPOSE, she'd just chosen for her far-famed book club. And naturally, it's all online.[9]

By way of introduction, Oprah sez, "Welcome to our very first live world-wide interactive event. We are here tonight breaking new ground. Nothing like this has ever been attempted before. Right now, you all are online with me from every corner on our planet."

I dunno. The never-been-attempted claim could be challenged by any number of hot girls with web cams. If anything, Oprah seems more than just fashionably late to the web party.

And Tolle sez, "It's everybody's responsibility to become aware of their conditioned mental processes: how you react in everyday situations, what kind of thoughts go through your head. It's good to not amplify the negativity that you see around you in the world by reacting to it."

Um...you mean like reading this sort of thing and reminding anyone who

will listen that your rap is basically recycling the same sort of Mystic Orientalia
that Emerson and Thoreau unleashed upon an unsuspecting Puritan New
England (when they weren't sowing beans and cheesing off the Harvard fac-
ulty), or Madame Blavatsky with her Theosophy (when she wasn't pontificating
on the Lost Continent of Atlantis), or Rudolf Steiner, founder of the Waldorf
Schools, with his stories about how Buddha taught the Mars Men (I wish I
were kidding), or...well...you mean like that?

Contra Tolle, I think it's not only "good" to amplify the negativity we see
around us by vaguely "reacting" to it, but more specifically by deconstructing,
dissecting, and calling it what it is. In this case, ten pounds of shit in a two-
pound bag. Is this what we've come to as a culture? Body by Barbie, epistemol-
ogy by "What the Bleep"?

The world of "spirit" and the world of politics have looped back on themselves like the two ends of that mythic—excuse me, *archetypal*—snake, the
ouroboric band encircling far more than those wagons of yore pushing west. It
is from the oil of this particular snake that we get Manifest Destiny, which
derives from the same celestial neighborhood as the Invisible Hand. Both represent God's holy will for Americans to be free, white, and twenty-one. Roughly
speaking. Or at least filthy rich. In brief, Manifest Destiny is the uniquely
American impulse to ingest every possible scrap of other people's sovereign territory in a God-frenzied desire for self-aggrandizement. This is sometimes also
referred to as "The Law of Attraction," or "Colonialism."

Much later, this colonial impulse will gather to itself other appellations:
globalization, the global economy, global marketing, global e-commerce, the
global Internet.

Manifest Destiny was big in the nineteenth century. For some weird reason—probably Indians who haven't yet set up casinos—politicians don't like to
talk about it so much these days. But it was especially popular back at the time
of the Louisiana Purchase (a.k.a. the French and Indian War), the Gadsden
Purchase (a.k.a. the Mexican-American War), the Oregon Treaty (a.k.a. the
"Fifty-four Forty or Fight" War), and the Southern Accommodation (a.k.a. the
Civil War). As you can see, war is a main ingredient of manifest destination.

But a funny thing happened when this basic instinct hit the Pacific Ocean. No amount of distinous manifesting seemed to have any effect on the water, which resisted the predations of all but stoned hipsters on surf boards. Until, that is, in the magic 1960s, Aldous Huxley, Alan Watts, Michael Murphy et al. had a breakthrough vision of the American future: Why not go vertical and colonize the *Mind*? Thus was Esalen Institute born, hard by the ancient Indian hot tubs and vast marijuana plantations of Big Sur. Fritz Perls served as the resident model for R. Crumb's Mr. Natural, while Abraham Maslow disguised his long-standing fascination with dominance by rebranding it as self-esteem.[10] And just about anybody who was everybody—goo-goo-ga-joob—Was There Then.

As P. T. Barnum said, there's one born every minute. Man, what a *market*! Conversations optional.

If politicians don't like to talk about Manifest Destiny, spiritual (but heavens, not *religious*) types show no such reservations. Exhibit A would have to be Wayne Dyer's best-selling MANIFEST YOUR DESTINY: THE NINE SPIRITUAL PRINCIPLES FOR GETTING EVERYTHING YOU WANT. Dyer is in a long line of American New Thought practitioners—a line that included Mary Baker Eddy and her Christian Science (sci-fi writer L. Ron Hubbard had lots of inspiration for Scientology). Dyer's general thrust in MANIFEST YOUR DESTINY is that you can be what you want to be—on Cloud 9. This is sometimes also referred to as "The Law of Attraction" or "Wishful Thinking." When invoked by affluent whites (i.e., not the Temptations, who were talking about heroin addiction), it is often a symptom of aggravated spiritual greed syndrome (DSM-IV: ASGS).

Well, so what? What's all this—repo men, beat poets, interstate highways, Ike, Oprah, Eminem, Al and Tipper, mooseburgers, poodles, manifest density, and what the bleepers—got to do with anything? Skipping over the fact that the world is slated to end in 2012 when the Mayan calendar's Energizer Bunny runs down, let me give you an example a bit closer to home. In the run-up to the election, among the many places I looked for clues as to Obama's chances, I visited Gallup.com, the online arm of an outfit that dates back to 1935. Wikipedia tells us, "To ensure his independence and objectivity, Dr. Gallup resolved that he would undertake no polling that was paid for or sponsored in any way by special interest groups...a commitment that Gallup upholds to this day."[11]

Independent and objective. Sounds solid, right? Trustworthy. Then imagine my surprise on discovering that Deepak Chopra—best-selling author of such titles as THE SEVEN SPIRITUAL LAWS OF SUCCESS and QUANTUM HEALING—is a "senior scientist" with the company. This is an organization, keep in mind, whose reports on the leanings of the American electorate substantially shape the output of mainstream media. Why is he involved at all? His Gallup bio says, "Dr. Chopra is the world's leading living authority on 'connectedness.'"

Connectedness? WTF? I rabbit off to Google and search it up. At his publisher's site (Random House), I find a passage from Chopra's book THE SPONTANEOUS FULFILLMENT OF DESIRE: HARNESSING THE INFINITE POWER OF COINCIDENCE TO CREATE MIRACLES. Miracles. I see. I also see that "connectedness" is a subset of something he calls "synchrodestiny." I wonder whether "synchro" is a secret handshake sort of synonym for "manifest." But I read on. It seems that synchrodestiny "allows us to see meaning in the world, to understand the connectedness or synchronicity of all things, to choose the kind of life we want to live, and to fulfill our spiritual journey."

This is science? By whose definition is this even remotely proximate to the scientific ballpark? The equation of connectedness with synchronicity provides a clue. The latter is a concept developed by the psychologist Carl Jung—who had a lifelong love affair with the paranormal and all things esoteric and occult—and Wolfgang Pauli. While Pauli was a bona fide quantum physicist, the concept of synchronicity was based on Jung's analysis of his *dreams*. It is not accepted as part of the scientific canon. We are dealing here with a form of high-flown obfuscatory mysticism, not any sort of grounded, rational analysis.

But who can tell anymore? This kind of "spiritualized" self-help jargon masquerading as "independent and objective" thought has invaded every shelf in the major bookstores, online and off, from science to psychology, from medicine to business. In the case of business books—and websites by the million—much of this New Age blatherskite is sneaked in, Trojan Horse–style, in the form of "life coaching" or various wrinkles on so-called positive psychology.

Abraham Maslow is something of a father figure to both. His management concepts are collected in THE MASLOW BUSINESS READER and his "hierarchy of needs" is often cited in the literature of business management. PEAK: HOW GREAT COMPANIES GET THEIR MOJO FROM MASLOW, is a more recent hagiography. But few know the Maslow who wrote, in his 1987 book, MOTIVATION AND PERSONALITY, "The study of crippled, stunted, immature, and unhealthy specimens can yield only a cripple psychology and a cripple philosophy."

Specimens? That's an uncomfortably harsh repudiation of the mentally ill, but it should come as no big revelation. Maslow's primary mentor was Edward L. Thorndike, a prominent interwar advocate of eugenics. And it was to American eugenics that Nazi Germany looked for inspiration in the development of the Final Solution—an extermination project that began not with Jews or Gypsies but with patients in mental hospitals. Let's hope it was a peak experience for them.

If Maslow is counted as an inspiration for positive psychology (a course on which is the most highly attended class at Harvard these days), then Martin Seligman is its reigning monarch. His pop-psych book, AUTHENTIC HAPPINESS: USING THE NEW POSITIVE PSYCHOLOGY TO REALIZE YOUR POTENTIAL FOR LASTING FULFILLMENT (2002), is based on his earlier work documented in LEARNED HELPLESSNESS: A THEORY FOR THE AGE OF PERSONAL CONTROL (1993). Hold that thought: control. The central insight of the latter book derived from delivering painful electrical shocks to dogs until they no longer attempted to even resist this...well, let's call it what it was: torture. Unsurprisingly, it was on the basis of this "learned helplessness" research that the CIA invited Seligman to consult on techniques that ended up being used on human prisoners at Abu Ghraib and Guantanamo. This is well documented by Jane Mayer in her 2008 book, THE DARK SIDE: THE INSIDE STORY OF HOW THE WAR ON TERROR TURNED INTO A WAR ON AMERICAN IDEALS. She also discussed Seligman's involvement in a follow-up interview for HARPER'S.[13]

Yet business can't seem to get enough of the guy. For one of many instances, see Pepperdine University's MBA course: The Positive Psychology Approach to Goal Management.[14] It is also noteworthy that his work has been strongly supported by the Templeton Foundation, which even offers a "Martin E. P. Seligman Award."[15] The foundation's homepage banner reads, "Supporting Science—Investing in the Big Issues." For Templeton, the big issues clearly trend toward the transcendent. Among its "core themes" we find a "Prayer & Meditation" project, which "supports research to advance objective scientific understanding of prayer and meditation."[16] And Templeton has the bucks to enlist many credentialed academics into such "objective scientific" studies. Its annual Templeton Prize is larger than the Nobel.

Are we dealing with cutting-edge science here? With a new form of esoteric religion? Or with some hybrid brand of cognitive snake oil whose origins date back to the nineteenth century? My money is on the last option. Proving these connections to a fare-thee-well is beyond the scope of this chapter (I've always wanted to write that), but various aspects of the theme are greatly expanded on my Mystic Bourgeoisie blog, to which the reader is kindly referred (I've always wanted to write that too).

Meanwhile, whether it's America's long fascination with Higher Power, from Jonathan Edwards to AA, Maslow's preoccupation with dominance-as-self-esteem, Seligman's come-hither cajolery re: personal control or the imposed corporate culture of command and control we wrote about in THE CLUETRAIN MANIFESTO, if anything can be said to be counter-CLUETRAIN it is an infatuation with authority. In Chapter 1 of the book, I wrote:

> The Internet is inherently seditious. It undermines unthinking
> respect for centralized authority, whether that "authority" is the
> neatly homogenized voice of broadcast advertising or the smarmy
> rhetoric of the corporate annual report.

Ironically, authority seems to be making a comeback, using the very medium we naively thought might defeat it. How to explain this? I got valuable answers to this question from John Dean. If you were around for Watergate: yeah, *that* John Dean. His CONSERVATIVES WITHOUT CONSCIENCE (2006) unpacks psychologist Robert Altemeyer's groundbreaking research on authoritarianism. Thanks to Dean popularizing his work, Altemeyer has put the full text of his most recent book, THE AUTHORITARIANS, on the web.[17] Most of us think of authoritarians as despots and dictators, but Altemeyer says that the more common type is what he calls the "authoritarian follower." By their numbers and degree of willing submission, it is these followers who create the seductive opportunity for authoritarian leaders. It's a symbiotic, if twisted, relationship. Here's Altemeyer in his own words:

> Authoritarian followers usually support the established authorities in their society, such as government officials and traditional religious leaders. Such people have historically been the "proper" authorities in life, the time-honored, entitled, customary leaders, and that means a lot to most authoritarians. Psychologically these followers have personalities featuring:
>
> **1.** a high degree of submission to the established, legitimate authorities in their society;
>
> **2.** high levels of aggression in the name of their authorities; and
>
> **3.** a high level of conventionalism.[18]

Thank God we put "business as usual" behind us, eh? What has changed, I would offer, in the formula above, are the "traditional" and "time-honored" bits. Our culture has succeeded, virtually overnight, in elevating utter nonentities to the status of authority figures: those famous for being famous, those powerful by dint of usurping power, those seen as wise because they're wealthy, those revered as numinous for channeling the Easter Bunny.

If I were blogging this, I could show you—in graphic detail, as they say— some killer examples of the true believers and jive-ass jingoists to whom Altemeyer's criteria apply. But this isn't a blog and lawyers are *so* expensive. Maybe another time. For now, ten years after CLUETRAIN, I think I need to reach back much further for a closing argument. Here it is:

Don't follow leaders. Watch your parking meters.

Bob Dylan, "Subterranean Homesick Blues," 1965

Notes

1. http://www.saltana.org/1/docar/0307.htm.

2. http://www.whitehouse.gov/briefing_room/.

3. http://skepdic.com/ramtha.html.

4. http://mysticbourgeoisie.blogspot.com/2006/07/obedient-poodles-of-boulder.html.

5. http://www.time.com/time/magazine/article/0,9171,770412,00.html.

6. http://www.time.com/time/magazine/article/0,9171,761217,00.html.

7. http://www.pbs.org/mormons/themes/prohibition.html.

8. http://mysticbourgeoisie.blogspot.com/.

9. http://www.oprah.com/article/oprahsbookclub/anewearth/20080130_obc_webcast_download.

10. http://books.google.com/books?id=LJp8AQAACAAJ.

11. http://en.wikipedia.org/wiki/Gallup_Poll.

12. http://www.gallup.com/corporate/19324/Deepak-Chopra-MD.aspx.

13. http://harpers.org/archive/2008/07/hbc-90003234.

14. http://gbr.pepperdine.edu/053/goals.html.

15. http://www.templeton.org/prizes/martin_ep_seligman_award/.

16. http://www.templeton.org/funding_areas/core_themes/prayer_and_meditation/.

17. http://home.cc.umanitoba.ca/~altemey/.

18. http://member.shaw.ca/jeanaltemeyer/drbob/chapter1.pdf/.

In Defense of Optimism

DAVID WEINBERGER

OH, WE WERE FEELING OUR OATS IN 1999.

The Internet was a primal force, a hurricane aimed straight at the mightiest institutions in the world. It was going to sweep them away, while—quite un-hurricane-like—giving us the tools to refashion what those institutions had gotten so wrong.

The Internet was letting us get it right, attacking the inhuman alienation characteristic of institutions driven from the top. Every layer of management, every calculated message delivered to the mass market, every memo from every lawyer urging staying on the "safe side"...all that just phonied the business up. The Internet was a force of disintermediation, as we used to say. It's not just that it let us deal directly with the people building the stuff we wanted to buy. The Internet was disintermediating the sedimentary layers of increasing nonhumanity. Person-to-person is also human-to-human, with no institutional assholism in the picture.

It turns out, we were basically right. The Internet does indeed let us connect human-to-human. We do route around a-holes on the Net. Business has been changed, often in ways that are invisible to us because we take them for granted.

Basically right, but not entirely. Beyond the triumphalist overstatements (Thesis 74: "We are immune to advertising. Just forget it."), which at least had a rhetorical purpose, we were wrong about how long it was going to take to throw the rascals out. Speaking for myself (as we four authors always do), I naively underestimated how resourceful the incumbents would be in resisting the change.

Only three years ago, my friend Ethan Zuckerman and I drafted an article arguing that anonymity ought to remain the default on the Internet. Coalitions were gathering—some trying to keep us safe, others wanting to chill speech they don't like—to make it impossible for anyone to use the Internet without identifying herself in traceable ways. Our manuscript pointed to the very positive effects of keeping anonymity as the norm on the Web. We closed by spinning out what would have to happen to prevent a determined and somewhat technical evildoer from keeping his or her communications private. You'd have to lock down computers so only officially approved applications could be loaded, you'd have to register programming tools as if they were guns, you'd have to pass and enforce laws requiring Internet cafés to demand strong identification from users, you'd have to monitor all traffic using deep-packet inspection that looks for suspicious words in every communication, and so on. It was, we thought, a catalog of absurdities, requiring such significant changes in hardware, software, laws, and culture that no one would think it possible.

Ethan and I put the article aside for no particular reason. I went back to it recently and discovered that our list of ridiculous conditions were now being considered and sometimes implemented. The Net that the CLUETRAIN celebrated is in danger.

The risk is serious. It threatens what is most valuable and transformative about the Internet: the way it lets us create new ways of connecting, one to some, many to many, and in ménage à whatevers we haven't yet invented. Of course, that threat—promise—of transformation is exactly what spurs incumbent institutions to oppose it.

So, let's first get good and depressed by looking at the threats that CLUETRAIN did not foresee back in the days when oats were oats and the

Internet was our inevitable future. And then we can talk about our duty to be optimistic.

Part 1: Threats of, by, and for the People

THE ROSTER OF THREATS IS DAUNTING, INCLUDING SOME OF THE most powerful institutional forces in our culture. Gulp.

1. The Access Providers

The folks who have been charged with providing the Internet to the citizenry of the United States of America love the Internet—the same way a real estate developer loves an old-growth forest with a mountain view.

A few years ago I was invited to give a talk at a cable industry conference. They scheduled me for an afternoon keynote. That morning, I used my name badge to attend a panel of CEOs and senior managers. As they waxed enthusiastic about "the challenges ahead," my spirits sank. I had known they want to turn the Internet into cable TV, but I hadn't realized the unassailable coherence of their point of view. To them, the Internet is a new way to deliver content and services. It makes more content available, the content is available on demand, and new services, such as Internet telephony, can be layered in as they are invented. But the Internet has one huge disadvantage from the viewpoint of these very serious men: The "customer experience" sucks. The programming isn't anywhere near high-def quality, telephone calls can sound like the cable is made out of connected kazoos, and everything is too hard to do. Therefore the cable industry's mandate has all the force of a syllogism: Make the Internet as easy to use and as high-quality as cable TV.

This makes perfect sense given two suppositions. First, you have to assume that the Internet is primarily there to deliver content. Second, you have to be willing to sacrifice everything else to that goal. Or, put differently, this makes perfect sense if you're willing to be entirely wrong about the Internet.

Every meme-flogging blogger and every pixel-pushing artist would agree that the Net is a way to deliver content. The problem is with the word "primarily."

The very characteristic that distinguishes the Internet from all other media is that it isn't primarily for any particular use. It just moves bits from any A to any B. Some of those bits are content of the sort we get on TV, but some are muzzle flashes in an online game, some are EKGs, and, yes, many are conversations. The Internet is for anything we can do by putting bits in motion. Once we decide that the Internet is primarily for one purpose, we then rationally have to give precedence and priority to that purpose. As Internet architect David Reed says, optimizing for one purpose means deoptimizing for all others. So, there goes user choice and there goes the ability to innovate tomorrow's new service on an equal footing with today's.

If you think the Internet is primarily for delivering professional video content, you speed up video bits, which means slowing down YouTube bits, as well as every other nonprofessional video. If you think your customers want an easy-to-use Internet, you herd them into your own portal onto the Net, which means your company decides which services are too hard for your poor dumb customers, which means the open market for good ideas—some of which may require more intelligence than it takes to press a TV remote—has been skewed in favor of the lowest and commonest of the denominators. If you have decided that the Net's real purpose is to keep us fat 'n' happy on our couches, bathed in the warm glow of messages from our sponsors, you give your customers the ability to download bits many times faster than they can upload them. That tells you exactly how conversational they think the Net is: not very.

The panel of well-groomed middle-aged men I was listening to that day didn't think they hated the Internet. They thought they had gotten past their initial denial and had gone all the way to enthusiasm. The Internet is, in their word of choice, an "opportunity." They just have to clear-cut the forest, pull up the stumps, rototill the remaining land, lay down some sod, and they'll have the suburban box of which every American dreams.

Let's hear it for the stumps.

2. The Content Cartel and Its Enablers

If you think the value of the Internet is the value of the content on it, then you likely think of copyright as your warrior king. In this way copyright becomes the first among all rights, and thus does it twist our culture around its bony fingers.

Of course copyright serves a purpose: "The Congress shall have Power to promote the Progress of Science and useful Arts, by securing for limited Times to Authors and Inventors the exclusive Right to their respective Writings and Discoveries." Our forefathers wrote that in beautiful script in Article I, Section 8, Clause 8 of our Constitution. Copyright exists to provide an incentive to creators to continue creating. Yet no one can argue that a writer would be too discouraged to write a book if copyright didn't deliver a royalty check to her graveside address a full seventy years after she's put up her personal RIP® sign. Copyright's current death+70 term thus has nothing to do with incentives. Rather, copyright now exists in its current form to prop up industries that used to benefit from the natural scarcity of material goods. Now that creative works are plentiful, copyright tries through law to reinstate a scarcity that technology has obviated and that the market has repudiated.

The evidence of this is all around us. Computing systems lock us out of reusing content even in quite legal ways by baking digital rights management (DRM) capabilities into their operating systems and hardware. One provider of Internet access—AT&T—has announced it will patrol its network for what it considers to be infringing material. Apple iTunes has required you to hack your own computer to transfer songs from one machine to another, and Apple's iStore only lets you install iPhone software that it has approved.

None of this is surprising. The entertainment industry now includes not only the moviemakers but the network providers and the software and hardware manufacturers, and they all have an interest in keeping their content scarce, even if it means that culture now runs into a dead-end, and our progress in science and the lively arts is throttled so the mainstream can continue to dazzle us with its retreads and safe bets. If we took as our goal the maximum flowering of culture, rather than protecting the interests of a tiny handful of producers and publishers, we could have a world full of music and creativity.

3. Government

When governments think about security, the Internet looms as an enabler of their worst fears. Against the image of a city getting nuked, the arguments in favor of an open Internet have no purchase. What good does it do to be able to chat, update your Facebook page, and enter market conversations if you're a wet spot in a crater?

It's a tough argument to counter because, taken in itself, it's entirely right. There are many bad actors out there, from the perpetrators of botnets to actual goddamn terrorists. Preferring to be vaporized rather than lose your right to download porn automatically classifies you at the far end of the spectrum of obsessive masturbators. The fear of bad actors, however, is used by police states to justify every conceivable limitation of rights. And in fact we are not facing a choice between watching our children get blown up or limiting net freedoms. Rather, we are facing a choice between a risk of attack and our freedom to speak, read, and connect. (Also: to masturbate.) That's a harder balance to strike, but we in Western democracies have tended to give much greater weight to freedom than to fear.

One reason we've favored freedom is that maintaining a totalitarian regime is a pain in the butt. You have to recruit spies to infiltrate neighborhood groups, you have to steam open letters, on every street corner you have to station men with Uzis in their hands and death symbols on their caps. Real-world totalitarianism is a full-time job for a regime. But on the net you can get the rudiments of a police state up and running in just about no time, and the bulk of the work can be done automagically! Centralize the ISPs. Require people to log in with their official state identification number. Do deep-packet inspection to look for transmissions using the sorts of words used by enemies of the state and music freeloaders. Investigate anyone who encrypts messages. Totalitarians get much more bang for their buck online.

So, the new levels of threat have surfaced just as total surveillance has become far easier. To complicate the matter further, we have a new sense of "privacy" that defaults to making things public in the hope that with so much available, no one will notice that photo of me half naked and drunk down on Fisherman's Wharf. So, what the hell. Let the Robot Squad check my text messages for suspicious phrases if the alternative is being indirectly responsible for the next terrorist attack.

I personally am not an absolutist about this. I would rather not die, and I feel some obligation to my children to prevent them from being flambéed. There are some government intrusions that don't bother me as much as they bother my fellow travelers. But I do fear the slippery slope of trading freedom for security. Unlike most slippery slopes—if we let homos marry, next thing you know we'll be letting people marry their cats, to paraphrase Rick Santorum—this slope we know historically is genuinely slippery. It's a short distance between deep-packet inspection and putting "suspicious" people on the no fly-list. And there's no distance at all between putting people on lists and chilling free speech.

Freedom, we seem to think, is no longer worth the risk.

4. The People Who Sell Us Stuff

Doc's phrase "Markets are conversations" is so powerful that it tends to wipe the mind clean of any other idea in THE CLUETRAIN MANIFESTO. And I think the phrase is right: Markets *are* conversations.

With that phrase in mind, newly Web-savvy marketers have marched proudly forward and made a mess of things.

Some have done so with good intentions. They believe what they read in CLUETRAIN or elsewhere. They understand that networked markets are usually the best source of information about products and services. They feel humbled and want to start talking in human voices. But they don't know how to. They don't have a sense for the ethos of the Web conversations they're jumping into. They don't have the patience to lurk and learn. So, they embarrass themselves. It's painful to watch, say, Wrigley's Gum introduce a "blog" in which two made-up characters express their unhealthy obsession with chewable candy. Or there's the fun of watching Doritos jump on the Web 2.0 bandwagon by offering Superbowl airtime to the best "user-generated" commercial, only to pick an ad as slick as one churned out by a professional shop...and then to find out that the ad was in fact created by an advertising professional. Such farcical behavior mirrors what some comedian once said about airline food: It's as if aliens saw pictures of food and knew nothing else about it.

But some marketers do not have good intentions. They are not stumbling on a path that overall is righteous and commendable. They see something of value and wish to corrupt it. They pay people to wax enthusiastic about their products in comments around the Web, without acknowledging the payment. They set up phony blogs. They take advantage of the anonymity that protects genuinely risky speech. They erode the trust that enables us to converse, and for that they should rot in some outer circle of hell.

Sure, some of them get caught at it. But by definition, we don't know how many get away with it. Just as the open Internet paradoxically enables governments to spy on citizens' online behavior, market conversations can provide cover for lying sacks of ordure. Conversations assume trust and thus are quite corruptible. And for that reason, predatory, opaque marketing is a threat to the heart of the Net's value as how we connect with one another.

5. The Digital Divide

The fact that access to the Internet falls out generally along the old tired lines of economic class and ethnicity is unfortunate, sad, and predictable. This failure of the Internet is due to its being nestled within a real and wildly unfair world. In fact, the link between atoms and unfairness is well-nigh unbreakable.

So, we have a digital divide that prevents the Internet from being all it can be. Worse, we will always have a digital divide of some sort. Even if (when) everyone is connected, some will be more connected than others. Some will have fast broadband and big screens, and others will be using those people's castoffs. Some will have the skills required—the media literacy, the social skills, even the words-per-minute typing skills—and some will not.

Still, there's an essential difference between having a slow connection and having no connection. Fortunately, of all the various divides, the digital one is one of the easier ones to overcome. We'll get everyone on the planet a handset, some bandwidth, and some basic Net skills well before we've ended the economic, ethnic, or gender divides.

6. People

THE CLUETRAIN MANIFESTO painted a very positive picture of the Internet, to say the least. I'm okay with that. We wanted to explain what Netizens were so excited about, and what businesses insisting that the Net changed nothing was causing them to miss. Pushback was inevitable, especially if you're thinking of the Internet as primarily being about content, and pushback is entirely proper.

One of the strongest arguments against the transformative power of the Net has been that the Internet doesn't make us more open to diversity. In fact, the argument goes, the Net is enticing us to hunker down with people who believe exactly the same things as we do, thus battening down our beliefs rather than opening them up. If that's right, then our own discomfort with difference is a serious threat to the value of the Internet.

What would it take to settle the argument? The maps of links among sites do indeed show that some sites are hubs and have many more links than others. When it comes to political sites, we seem to cluster around a relative handful of partisan places. But link maps don't tell the whole story. We don't know people's actual reading habits. Do the people who hang out in these clusters *only* hang out there? Does the homogeneity of the clustering vary by topic? Does it vary by political stance? By demographic data? By the urgency or freshness of the issue?

Let's say we answer those questions. Let's say we know, on average, how many different sorts of sites people traverse, what sorts of mailing lists they're on, and who they hang out with on their social networking sites. We still wouldn't know if the Internet is making us more or less diverse because we don't have a baseline against which to compare this data. Since in the United States, newspaper op-eds run the entire spectrum of political thought from forest green to light forest green, are we sure that our daily papers are more diverse than these Web clusters? Do people read their newspapers end to end, or do they skip the articles and editorials they don't care for and focus on those they agree with? If newspapers form our baseline, how do we figure out what that baseline actually is?

Maybe we should compare the Internet not to newspapers but to the conversations we have in the real world. How diverse are they? If we could listen in, we'd probably find that people talk with those who are more like them than unlike. Indeed, conversation requires a broad base of agreement from which we then discuss relatively minor differences. Conversation isn't usually about finding the truth. It's a social activity and a way of building social relationships. Those who despair that we are living in Net bunkers may well have an unrealistic view of the role of conversation.

So, to decide whether the Net is making us more or less open to diversity, we'd have to answer an ever deepening series of questions that begin with the empirical (What actually is the spread of Net experiences when it comes to encountering the unfamiliar?) and descend into the sociophilosophical (What is the role of conversation?).

When the bottom drops out of a question as simple as this, we should worry that there's something wrong with the question itself. But assume for the moment that the cynics have a legitimate concern, for they do. If they are right that the Net hardens our hearts and our minds, our possibly innate preference for similarity over difference would make human nature itself the most powerful enemy of the Internet ever.

Oh, great.

Part 2: Rainbows, Kitty Cats, and Cyberutopianism

HERE'S THE EXECUTIVE SUMMARY OF PART 1: THE BEAUTIFUL PROMISE of the Internet is heading for the crapper. Like my ex-hippie generation that twice elected George Bush, we're about to blow it. Big time.

But here's the executive summary of Part 2: Don't worry. Be happy! Among the three major attitudes toward the Internet, choose the happiest. It's not just more fun. It's your duty. That would make you a cyber utopian, much like the authors of THE CLUETRAIN MANIFESTO in 1999. This tribe points to the ways in which the Net has changed basic assumptions about how we live together, removing old obstacles and enabling shiny new possibilities. Of course, you could become a cyber dystopian who agrees that the Net is having a major effect on our lives, but you think the effect is detrimental. Finally, you could become a cyber realist who pooh-poohs the magnitude of effect that has so awed the utopians and the dystopians. As a cyber realist, you would say that the possibilities and limitations of the Net are really not so much greater than those of any other major communications medium.

So, who's right?

Consider this hypothetical exchange:

> Realist: **You say that the Web will transform politics. But politics is the same as it ever was.**
>
> Utopian: **Just wait.**

This is indeed one of the two basic blocking tactics used by cyber utopians: The changes are so important that they will take a while to arrive. Here's an example of a second utopian tactic at work:

> Realist: **You say that the Web will transform business, but business is the same as it ever was.**
>
> Utopian: **Not at all! For example, email has transformed meetings, but we're so used to the change that we don't even recognize it.**

To this, the cyber realist has a number of responses: denying that the changes are real, that they are important, or that they are due to the Web.

When a dystopian points to a bad effect of the Web, the utopian denies the claim's truth, inevitability, or importance:

> Dystopian: **The Web has made pornography available to every schoolchild!**
>
> Utopian: **It is the responsibility of parents to make sure their kids are using child-safe filters. Besides, viewing pornography may weaken our unhealthy antisex attitudes. Besides, greater access to porn is just one effect of the Web; it's brought greater access to literature, art, science...**

But surely someone has to be right about whether the Internet sucks or is our savior. How about gathering some facts? Fine. Which ones? Do we take a company caught paying customers to write positive blog posts as an aberration or as typical? Do we focus on the presence of online child porn or online assistance for new parents? Do we put forward forums where grieving parents support one another or flame fests where cretins drive each other to new depths?

Facts are of little avail because utopians, dystopians, and realists have different views of history, and the framing of history also frames facts. Utopians and dystopians think the Web has uncanny power because they are closet McLuhanites who think media transform institutions and even consciousness. Realists feel the inertial weight of existing institutions and social structures, and thus tend to think any changes the Internet brings will be slow and minimal. After all, say the realists, nothing as trivial as HTML will change the fact that corporations are firmly in control, and that much of the world lives in helpless poverty. One's view of history frames the facts one attends to. Facts are therefore unlikely to upend one's historical frame.

Indeed, these three stances are political, not factual. Utopians want to excite us about the future possibilities because they want policies that will keep the Internet an open field for bottom-up innovation. Dystopians want to warn us of the dangers of the Web so we can create policies and practices that will mitigate those dangers. Realists want to clear away false promises so people don't fritter away time on airy-fairy projects that will lead nowhere. Also, they'd like the blowhard utopians to just shut up for a while.

From these positions flow practical policies. If the cyber utopians are right that the Internet is transformative in an overall positive way—the most basic premise of the original CLUETRAIN—then it's thus morally incumbent on us to provide widespread access and training to as much of the world as possible, focusing on the disadvantaged. If the dystopians are right, we should put the brakes on access and immediately put in place whatever safeguards we can. If the realists are right, then we ought to make tactical adjustments and ignore the hyperventilations of the utopians and dystopians. If the Web is transforming business, for better (utopian) or for worse (dystopian), then businesses need to alter their strategic plans. If the Web is merely one more way information travels (realist), then businesses should learn the techniques for taking advantage of the new channel but do nothing more.

If we cannot choose among these three attitudes based solely on facts, then how do we choose? I suggest that we pick the one with the best outcome. On those grounds, dystopianism is out of luck. It's fear-based and depressing. How are you going to run a business with Eeyore and Chicken Little as your role models? That leaves utopianism and realism.

Realism has the better name. It sounds like it has to be more in touch with reality. But what realist would have thought that Wikipedia was worth the time or investment? What realist would have said, "Yeah, getting a disorganized mob to build an operating system to compete with Microsoft is a terrifically practical idea"? What realist would have said in 1992, "I've seen Tim Berners-Lee's plan for a worldwide Web, and, realistically, in ten years there will be a billion people on it and hundreds of billions of pages"? A premature realism would have killed the Web. Who knows what earth-shaking ideas are right now being laughed off the Net by confident realists? Realism just isn't ambitious enough.

And for exactly the same reasons, utopianism is the Net's ally. Realists are right more often than utopians are, because most ideas are bad. But without utopianism, the ideas that are so good that they change the world would never get the air they need. Utopians have made more of the Internet than any reasonable person could have expected. Utopians rule.

Of course we need realists. Innovation requires the realism that keeps us from wasting time on the impossible. Realists keep us from running down dead-ends longer than we need to, and from getting into feedback loops that distort the innovation process. At the nittier-grittier levels of provisioning Internet services, hard-core realists with bits under their fingernails are a godsend. For all that and more, we should thank and encourage the realists. But realists are not enough. Besides, realism already has a good name; you don't have to convince anyone that realism is a good thing. Utopianism, on the other hand, has a bad name. We need to stand by our utopians so they can drive us into the impossible future. We need lots and lots of them. There is so much to yet to invent.

Yes, I admit the whole argument is essentially silly. We don't get to choose whether we're realists, optimists, or pessimists, any more than we get to choose if we're born in Albany or Beijing. We're pretty much stuck with who we are. We're also stuck in a culture we didn't choose to be born into, a family that raised us, a language that divvies up and links the world one way and not another, experiences that happen to us whether we like them or not, all on a planet that turns out to be good for growing delicious plants but that has the occasional lethal crevasse. That's the sort of stuck we are.

So, I'm not really arguing that 90 percent of us ought to become cyber utopians and 10 percent should become realists, or whatever the optimal mix may be. Rather, the implicit and basic utopianism of THE CLUETRAIN MANIFESTO and, far more importantly, of the hundreds of millions of people on the Internet, was helpful, important, and, as Martha herself would say, A Good Thing. It is not yet time to let a forlorn realism reign.

Part 3: What to Do

SO, WITH OUR NEW SENSE OF OPTIMISM IN CHARGE, IN PART 3 LET'S look back at the threats adumbrated in Part 1.

Those threats are real. I underestimated them in my contributions to THE CLUETRAIN MANIFESTO ten years ago. But the main lesson I've learned—or, at least I think it's a lesson and not just my delusion du jour—is not that cyber utopianism is wrong. It's that the values of the Internet won't become real without our help.

Let's revisit each of the threats from the first section.

1. The Access Providers

It wouldn't be all that hard to keep the access providers from being able to decide that content—especially their content—is what the Internet is *really* for. Net neutrality legislation would forbid them from giving their content and services precedence over anyone else's. Of course, the providers have a history of simply ignoring laws and policies that don't suit them. So we could, for example, once again require Net access providers to give competitors access to their lines at wholesale prices. We'd once again have thousands of ISPs outdoing themselves to offer us competitive deals.

Or we could decide that access to the Internet provides such an important advantage—economically, educationally, socially, politically—that the federal government has to step in to build national bitways, just as in the 1950s it built national highways. There are lots of potential configurations for accomplishing this, but all of them implicitly acknowledge that the way we're doing it now is not serving our national weal.

With a new administration just having been elected as I write this (November 2008), perhaps we will make some rapid progress. Change, hope, and all that.

2. The Content Cartel and Its Enablers

There's no denying we want high-quality professional content all over the Internet. And it'd be fair to criticize THE CLUETRAIN MANIFESTO for overemphasizing the importance of conversation over content. Not every good idea on the Internet is a conversation. (We never said it was.) But most often content on the Net comes wrapped in conversation. Consider the hundreds of millions of people who are giving up TV time to watch amateur YouTubes. A YouTube video is content, but YouTube without the ability to comment and, more importantly, to recommend videos to others wouldn't be YouTube. Or take the millions of people who watch credit card–size versions of broadcast TV on their laptops while they talk in real time with others watching the same little windows. If you haven't watched a presidential debate that way, you will; we are so much more interesting than Wolf Blitzer and the Best Political Team on Television.

Talking about—and over, around, and through—content is one way we appropriate it. It's how we absorb it. That's why digital rights management techniques hold back culture. They keep content at an arm's distance...and that arm is holding a legal gun. We thus can't absorb it or make it culturally our own. That's why there is no market demand for DRM. It's always imposed on us. We want to be able to use the stuff we buy, transfer it from our iPod to our laptop to our DVR. We want to be able to do remixes that enhance or ridicule what the mass market is giving us.

It's confusing and difficult because this isn't a struggle over property, like the owner of a beach house who tries to get an injunction to keep teenagers from swimming there. This is an ontological struggle: the nature of content as property is changing before our eyes, all part of an even larger rejiggering of privacy, culture, and collaboration. Folk art is in resurgence but via a medium more mass than any we've seen so far. There is the possibility of soaking the earth with music we make, borrow, and reuse. Meanwhile, we are stuck in economic and legal systems that work well for atoms and not so well for connected bits.

There are signs of movement away from locked-down machines and locked-down content, although there are simultaneous movements in the other direction. Perhaps someday soon we'll even get the new conversation about copyright that we so desperately need. In the meantime, we should take heart from the new sense on the Web that sharing is the norm, and that collaborating is a natural act.

3. Governments

The access providers and content cartel are looking out for their own interests, but the government's security forces are looking out for ours. If they say they need users to sign in with an ID number and a fingerprint, or if they say that they cannot tolerate encrypted communications, politicians are likely to cave in so they can't be held responsible for the nonprevention of the next attack. That's why the price of freedom is eternal vigilance. Eternal, constant, grinding, scary vigilance.

This is going to be a long, difficult struggle. I believe we'll only succeed if we supporters of openness are willing to lose some of our old instincts, just as we're asking the security forces of the world to lose some of theirs. When machines can process our packets, the chief problem is not that another person will see us naked but that the machines are really stupid, so innocent people may be threatened, and speech overall is chilled. One solution is not to totally ban the machine surveillance but to tightly control what happens when a machine thinks it's found something suspicious.

I don't know what the solution is. But we need to be prepared to engage in a lifelong struggle for our open Internet.

4. Marketers

The original CLUETRAIN told us what we need to do to prevent marketers from subverting our conversations: We need to talk with one another. We need to trust one another more than we trust them. We need to continue to evolve mechanisms for exposing conversational predators. We need to continue to innovate trust mechanisms that are appropriate to the nature of the engagement—stricter if money is changing hands, looser if people are just blabbin' about stuff.

The good news is that lots of marketers have already figured out that conversational crime doesn't pay. For example, the Word of Mouth Marketing Association started itself up with a strong code of ethics, and they've disciplined members who have violated it.

We've made a good start. We need to continue being wary so long as the temptation to lie is among us. Yes, I'm talking about forever.

5. The Digital Divide

We have to try everything we can think of, everything that makes sense, and a bunch of things that seem to make no sense at all. Governments, municipalities, philanthropies, private industry. Everything. Every form of connection, from laptops, to $100 laptops, to mobile phones, to nano paint. Whatever it takes to get people connected. And then whatever it takes to get people to learn to get—and make—what they want from their new connectedness.

Some of what we try will work. Much will fail. What works will never work well enough. But it will have to be good enough.

6. People

Fortunately we don't have to decide whether the Internet is making us more or less comfortable with diversity to know what to do about it. No matter how much the Internet is making us more open to differences, it is not enough and it will never be enough. Inertia is on the side of sticking with what—and who—you know. So we have to create sites like GlobalVoicesOnline.org to entice Americans to listen to voices from around the world. We have to remember to pass around links to ideas we don't agree with. We have to restrain ourselves from passing too quickly over what we don't immediately appreciate.

I still think the Internet overall increases our appreciation of difference, but I now think that it does not increase it enough *by itself*. We have to work on this, and we will always have to work on this.

The solution to each of these six threats is: Work. Struggle. Fight. Forever. For that, optimism is required. And the "forever" part means we have to go beyond optimism all the way to hope.

That's different than what I thought when we were writing CLUETRAIN. In that first edition, we seemed to think that although we wouldn't capture Fort Business without a fight, the Internet was a force invincible and inevitable. We were wrong about the "invincible" and "inevitable" part, but I still believe that history is on the side of the Internet, for two reasons.

First, each of the threats comes from the attempt to maintain the old scarcities in the face of the overwhelming abundance characteristic of the Internet. Access providers want to limit access so they can charge more for it; content providers use copyright to impose an artificial scarcity on their goods; governments want to limit access to thwart dangerous actors; marketers want to control what we know about their products; and our own nature seems to want to limit us to that with which we are already familiar. Ten years after CLUETRAIN, we know the Internet won't simply route around those limitations. But routing around isn't the only tactic. Floodwaters don't route around the fences in the fields they overwhelm. Abundance has its own tactics, for it turns out that control doesn't scale. The curve of the cost of control rises faster than the curve of growth. There's hope that there is a point at which abundance eventually outspends control.

Second, my most fundamental belief about the Net, one at the heart of what I've thought and written about since I first visited a single-column, plain text Web page in the ancient Mosaic browser, is that the Web allows a fuller, more natural expression of who we are than mass culture does. At bottom we are creatures who share a world with others we care about. There is no turning back from sharing and caring. Mr. Rogers had it right. And with hyperlinks, we have a technology well suited to our best nature.

I was an optimist in 1999. Ten years later, I am an optimist who feels he needs to defend optimism...and an optimist who believes we have to fight for the inevitable.

Internet Apocalypso

CHRISTOPHER LOCKE

you set my desire . . .
I trip through your wires

U2

Premature Burial

WE DIE.

You will never hear those words spoken in a television ad. Yet this central fact of human existence colors our world and how we perceive ourselves within it. "Life is too short," we say, and it is. Too short for office politics, for busywork and pointless paper chases, for jumping through hoops and covering our asses, for trying to please, to not offend, for constantly struggling to achieve some ever-receding definition of success. Too short as well for worrying whether we bought the right suit, the right breakfast cereal, the right laptop computer, the right brand of underarm deodorant.

Life is too short because we die. Alone with ourselves, we sometimes stop to wonder what's important, really. Our kids, our friends, our lovers, our losses? Things change, and change is often painful. People get "downsized," move away, the old neighborhood isn't what it used to be. Children get sick, get better, get bored, get on our nerves. They grow up hearing news of a world more frightening than anything in ancient fairy tales. The wicked witch won't really push you into the oven, honey, but watch out for AK-47s at recess.

Amazingly, we learn to live with it. Human beings are incredibly resilient. We know it's all temporary, that we can't freeze the good times or hold back the bad. We roll with the punches, regroup, rebuild, pick up the pieces, take another shot. We come to understand that life is just like that. And this seemingly simple understanding is the seed of a profound wisdom.

It is also the source of a deep hunger that pervades modern life—a longing for something entirely different from the reality reinforced by everyday experience. We long for more connection between what we do for a living and what we genuinely care about, for work that's more than clock-watching drudgery. We long for release from anonymity, to be seen as who we feel ourselves to be rather than as the sum of abstract metrics and parameters. We long to be part of a world that makes sense rather than accept the accidental alienation imposed by market forces too large to grasp, to even contemplate.

And this longing is not mere wistful nostalgia, not just some unreconstructed adolescent dream. It is living evidence of heart, of what makes us most human.

But companies don't like us human. They leverage our longing for their own ends. If we feel inadequate, there's a product that will fill the hole, a bit of fetishistic magic that will make us complete. Perhaps a new car would do the trick. Maybe a trip to the Caribbean or that new CD or a nice shiny set of Ginsu steak knives. Anything, everything, just get more stuff. Our role is to consume.

Of course, the new car alone is not enough. It must be made to represent something larger. Much larger. The blonde draped over the hood looks so much better than the old lady bitching about the dishes. Surely *she'd* understand our secret needs. And if we showed up with her at the big golf game, wouldn't the guys be impressed! Yeah, gotta get one-a those babies. This isn't about sex, it's about *power*—the greatest bait there ever was to seduce the powerless.

Or take it one slice closer to the bone. Leverage care. For the cost of a jar of peanut butter, you can be a Great Mom, the kind every kid would love to have. You can look out on your happy kids playing in that perfect suburban backyard and breathe a little sigh of contentment that life's so good, with not a wicked witch in sight. Just like on television.

We die. And there's more than one way to get it over with. Advertising has some serving suggestions for your premature burial.

Testing, Testing . . .

BUT WHAT'S THIS GOT TO DO WITH THE INTERNET? A LOT.

The Net grew like a weed between the cracks in the monolithic steel-and-glass empire of traditional commerce. It was technically obscure, impenetrable, populated by geeks and wizards, loners, misfits. When I started using the Internet, nobody gave a damn about it outside of a few big universities and the military-industrial complex they served. In fact, if you were outside that favored circle, you couldn't even log on. The idea that the Internet would someday constitute the world's largest marketplace would have been laughable if anyone was entertaining such delusions back then. I began entertaining them publicly in 1992 and the laughter was long and loud.

The Net grew and prospered largely because it was ignored. It worked by different rules than the rules of business. Market penetration wasn't interesting because there *was* no market—unless it was a market for new ideas. The Net was built by people who said things like: What if we try this? Nope. What if we try that? Nope. What if we try this other thing? Well, hot damn! Look at that!

One of the hottest damns was the World Wide Web. It came out of efforts to create electronic footnotes—references between academic papers on high-energy physics that maybe a few dozen people in the entire world could actually understand. That's why now, when you turn on your TV, you see www.haveanotherbeer.com.

Well, OK, a few things did happen in between. One of those things was that the Internet attracted millions. Many millions. The interesting question to ask is why. In the early 1990s, there was nothing like the Internet we take for granted today. Back then, the Net was primitive, daunting, uninviting. So what did we come for? And the answer is: each other.

The Internet became a place where people could talk to other people without constraint. Without filters or censorship or official sanction—and perhaps most significantly, without advertising. Another, noncommercial, culture began forming across this out-of-the-way collection of computer networks. Long before graphical user interfaces made the scene, the scene was populated by plain old boring ASCII: green phosphor text scrolling up screens at the glacial pace afforded by early modems. So where was the attraction in that?

The attraction was in speech, however mediated. In people talking, however slowly. And mostly, the attraction lay in the kinds of things they were saying. Never in history had so many had the chance to know what so many others were thinking on such a wide range of subjects. Slowly at first, a new kind of conversation was beginning to emerge, but it would achieve global reach with astonishing speed.

In the early days, the Internet was used almost exclusively for government-funded projects and the sort of communication that went along with such work. Here's the new program. It needs some work. There's a bug in the frimular module. Yawn.

But you know what they say about all work and no play. People began to play. Left to themselves, they always do. And the people building the Internet were pretty much left to themselves. They were creating the gameboard. No one else knew how the hell this thing worked, so no one could tell them what they could and couldn't do. They did whatever they liked. And one of the things they liked most was arguing.

Consider that these early denizens of the Net were, for the most part, young, brash, untrained in the intricate dance of corporate politics, and highly knowledgeable of their craft. In the prized and noble older sense of the term, they were hackers, and proud of it. Many, in their own assessment if not that of others, were net.gods—high priests of an arcane art very few even knew existed. When disagreements arose over serious matters—the correct use of quotation marks, say—they would join in battle like old Norse warriors:

"Jim, you are a complete idiot. Your code is so brain-damaged it won't even compile. Read a book, moron."

Today, we tend to think of "flaming" as a handful of people vociferously insulting each other online. A certain sense of finesse has largely been lost. In the olden days, a good flame war could go on for weeks or months, with hot invective flying around like rhetorical shrapnel. It was high art, high entertainment. Though tempers flared hot and professional bridges were sometimes irreparably burned, ultimately it was a game—a participatory sport in which the audience awarded points for felicitous disparagements, particularly well-worded putdowns, inspired squelches.

It was not a game, however, for the meek of heart. These engagements could be fierce. Even trying to separate the contestants could bring down a hail of sharp-tongued derision. Theories were floated and defended with extreme energy and enthusiasm, if not always with logical rigor. Opinions tended to run high on any given topic. Say you'd posted about your dog. And, look, you got a response! "Jim, you are a complete idiot. Your dog is so brain-damaged it won't even hunt...."

If you'd happened to see the first version of the comment to Jim, you might grin at the second. If not, your mileage might vary. But the point is not to extol flame wars, as amusing as some could be. Instead, it is to suggest a particular set of values that began to emerge in what linguists might call a well-bounded speech community. On the Net, you said what you meant and had better be ready to explain your position and how you'd arrived at it. Mouthing platitudes guaranteed that you would be challenged. Nothing was accepted at face value, or taken for granted. Everything was subject to question, revision, re-implementation, parody—whether it was an algorithm, a political philosophy, or, God help you, an advertisement.

While the outcome of these debates did not invariably constitute wisdom for the ages, the process by which they took place was honing a razor-sharp sense of collective potential. The conversation was not only engaging, interesting, exciting—it was effective. Tools and techniques emerged with a speed that broke all precedents. As would soon become obvious, the Net was a powerful multiplier for intellectual capital.

Waiting for Joe Six-Pack

A FEW YEARS AGO, YOU COULD MAKE AN INTERESTING distinction between people who thought there was something special about the Internet and those who saw it as no big deal. Now of course, everybody sees it as a big deal, mostly because of those weirdball IPOs and the overnight billionaires they've spawned. But I think the distinction is still valid. Most companies with Net-dot-dollar-signs in their eyes today are still missing the "something special" dimension.

Yahoo has already made it, financially speaking, but forms a good example nonetheless. Despite the funky hacker roots of the initial directory Yang and Filo built, Yahoo now describes itself as a "global media company," thus claiming a closer spiritual kinship with Disney and Murdoch than with the culture that originally put it on the map.

To this mindset, the Net is just an extension of preceding mass media, primarily television. The rhetoric it uses is freighted with the same crypto-religious marketing jargon that characterized broadcast: brand, market share, eyeballs, demographics. And guess what? It works. If nobody was getting rich off this stuff, you wouldn't hear about it.

It's the fast new companies that are reaping these monetary rewards. But guess what again? They're reaping them from an even faster market—one that, for the most part, has only discovered the Internet in the last year or so. The people who make up this new market naturally bring a lot of baggage from their previous experience of mass media. To someone who just got an AOL account last Christmas, I suppose a Web page looks like a v-e-r-y s-l-o-w TV show.

But this is where the something-special effect comes in. It is assumed in some quarters that if you missed the early days of Usenet and didn't use Lynx from a Unix command line, you missed the Magic of Internet Culture. I don't think so.

Sure those were very different days and there was a certain fervor—almost a fever—that was hard to mistake for sitcom fandom. But I think the Internet still has a radicalizing effect today, despite all the banner ads and promotional hype and you-may-already-be-a-winner sweepstakes.

The something special is what the Manifesto calls *voice*.

Imagine for a moment: millions of people sitting in their shuttered homes at night, bathed in that ghostly blue television aura. They're passive, yeah, but more than that: They're isolated from each other.

Now imagine another magic wire strung from house to house, hooking all these poor bastards up. They're still watching the same old crap. Then, during the touching love scene, some joker lobs an off-color aside—and everybody hears it. Whoa! What was that? People are rolling on the floor laughing. And it begins to happen so often, it gets abbreviated: ROTFL. The audience is suddenly connected to *itself*.

What was once The Show, the hypnotic focus and tee-vee advertising carrier wave, becomes in the context of the Internet a sort of reverse new-media McGuffin—an excuse to get together rather than an excuse not to. Think of Joel and the 'bots on MYSTERY SCIENCE THEATER 3000. The point is not to watch the film, but to outdo each other making fun of it.

And for such radically realigned purposes, some bloated corporate Web site can serve as a target every bit as well as GODZILLA, KING OF THE MONSTERS. As the remake trailer put it: Size *does* matter.

So here comes Joe Six-Pack onto AOL. What does he know about netliness? Nothing. Zilch. He has no cultural context whatsoever. But soon, very soon, what he hears is something he never heard in TV Land: people cracking up. "That ain't no laugh track neither," Joe is thinking and goes looking for the source of this strange, new, rather seductive sound.

So here's a little story problem for ya, class. If the Internet has 50 million people on it, and they're not all as dumb as they look, but the corporations trying to make a fast buck off their asses *are* as dumb as they look, how long before Joe is laughing as hard as everyone else?

The correct answer of course: not long at all. And as soon as he starts laughing, he's not Joe Six-Pack anymore. He's no longer part of some passive couch-potato target demographic. Because the Net connects people to *each other*, and impassions and empowers through those connections, the media dream of the Web as another acquiescent mass-consumer market is a figment and a fantasy.

The Internet is inherently seditious. It undermines unthinking respect for centralized authority, whether that "authority" is the neatly homogenized voice of broadcast advertising or the smarmy rhetoric of the corporate annual report.

And Internet technology has also threaded its way deep into the heart of Corporate Empire, where once upon a time, lockstep loyalty to the chairman's latest attempt at insight was no further away than the mimeograph machine. One memo from Mr. Big and everyone believed (or so Mr. Big liked to think).

No more. The same kind of seditious deconstruction that's being practiced on the Web today, just for the hell of it, is also seeping onto the company intranet. How many satires are floating around there, one wonders: of the latest hyperinflated restructuring plan, of the over-sincere cultural-sensitivity training sessions Human Resources made mandatory last week, of all the gibberish that passes for "management"—or *has* passed up until now.

Step back a frame or two. Zoom out. Isn't that weird? Workers and markets are speaking the same language! And they're both speaking it in the same shoot-from-the-hip, unedited, devil-take-the-hindmost style.

This conversation may be irreverent of eternal verities, but it's not all jokes. Whether in the marketplace or at work, people do have genuine, serious concerns. And we have something else as well: knowledge. Not the sort of boring, abstract knowledge that "Knowledge Management" wants to manage. No. The real thing. We have knowledge of what we do and how we do it—our craft— and it drives our voices; it's what we most like to talk about.

But this whole gamut of conversation, from infinite jest to point-specific expertise: Who needs it?

Companies need it. Without it they can't innovate, build consensus, or go to market. Markets need it. Without it they don't know what works and what doesn't; don't know why they should give a damn. Cultures need it. Without play and knowledge in equal measure, they begin to die. People get gloomy, anxious, and depressed. Eventually, the guns come out.

There are two new conversations going on today, both vibrant and exciting; both mediated by Internet technologies but having little to do with technology otherwise. Unfortunately, there's also a metaphorical firewall separating these conversations, and that wall is the traditional, conservative, fearful corporation.

So what is to be done? Easy: Burn down business-as-usual. Bulldoze it. Cordon off the area. Set up barricades. Cripple the tanks. Topple the statues of heroes too long dead into the street.

Sound familiar? You bet it does. And the message has been the same all along, from Paris in '68 to the Berlin Wall, from Warsaw to Tiananmen Square: Let the kids rock and roll!

So open the windows and turn up the volume. If the noise gets loud enough, maybe even CNN will cover.

From Ancient Markets to Global Networks

THIS MAY SEEM RABIDLY ANTIBUSINESS. IT'S NOT. Business is just a word for buying and selling things. In one way or another, we all rely on this commerce, both to get the things we want or need, and to afford them. We are alternately the workers who create products and services, and the customers who purchase them. There is nothing inherently wrong with this setup. Except when it becomes all of life. Except when life becomes secondary and subordinate. At the beginning of the twenty-first century, business so dominates all other aspects of our existence that it's hard to imagine it was ever otherwise. But it was. Imagine it.

Storylines

A few thousand years ago there was a marketplace. Never mind where. Traders returned from far seas with spices, silks, and precious, magical stones. Caravans arrived across burning deserts bringing dates and figs, snakes, parrots, monkeys, strange music, stranger tales. The marketplace was the heart of the city, the kernel, the hub, the *omphalos*. Like past and future, it stood at the crossroads. People woke early and went there for coffee and vegetables, eggs and wine, for pots and carpets, rings and necklaces, for toys and sweets, for love, for rope, for soap, for wagons and carts, for bleating goats and evil-tempered camels. They went there to look and listen and to marvel, to buy and be amused. But mostly they went to meet each other. And to talk.

In the market, language grew. Became bolder, more sophisticated. Leaped and sparked from mind to mind. Incited by curiosity and rapt attention, it took astounding risks that none had ever dared to contemplate, built whole civilizations from the ground up.

Markets are conversations. Trade routes pave the storylines. Across the millennia in between, the human voice is the music we have always listened for, and still best understand.

So what went wrong? From the perspective of corporations, many of which by the twentieth century had become bigger and far more powerful than ancient city-states, nothing went wrong. But things did change.

Commerce is a natural part of human life, but it has become increasingly unnatural over the intervening centuries, incrementally divorcing itself from the people on whom it most depends, whether workers or customers. While this change is in many ways understandable—huge factories took the place of village shops; the marketplace moved from the center of the town and came to depend on far-flung mercantile trade—the result has been to interpose a vast chasm between buyers and sellers.

By our own lifetimes, mass production and mass media had totally transformed this relationship, which came to be characterized by alienation and mystery. Exactly what relationship did producers and markets have to each other anymore? In attempting to answer this blind-man's-bluff question, market research became a billion-dollar industry.

Once an intrinsic part of the local community, commerce has evolved to become the primary force shaping the community of nations on a global scale. But because of its increasing divorce from the day-to-day concerns of real people, commerce has come to ignore the natural conversation that defines communities as human.

The slow pace of this historic change has made it seem unsurprising to many that people are now valued primarily for their capacity to consume, as targets for product pitches, as demographic abstractions. Few living in the so-called civilized world today can envision commerce as ever having been anything different. But much of the change happened in the century just passed.

Economies of Scale: Mo' Bigga Mo' Betta

The Internet is often seen as a unique phenomenon that only recently burst into the economic mainstream. But looking at the Net in strictly technological terms obscures its relationship to broader economic trends that were already well underway.

By the end of the nineteenth century, the United States was poised to become the prototypical mass market. It had vast natural resources, a fast-growing population, and a contiguous geography generally unbounded by tariff restrictions. Cheap iron coupled with a voracious appetite for industrial expansion enabled a railway system capable of cost-effectively delivering goods to nearly every part of a captive domestic market.

Given the high cost of entry into such enterprises, and without appreciable foreign competition, manufacturers cared little about product differentiation. Thus Henry Ford's attitude toward customer choice: "They can have any color they want as long as it's black." More than for his wit, Ford is remembered for designing the first high-volume automotive assembly lines. The more cars Ford could make, the lower the unit cost and the greater the margin of profit. These economies of scale led to enormous profits because they enabled selling a far cheaper product to a far wider market.

Ford was strongly influenced by Frederick Taylor and his theory of "scientific management." Taylor's time-and-motion metrics sought to bring regularity and predictability to bear on the increasingly detailed division of labor. Under such a regimen, previously holistic craft expertise rapidly degraded into the mindless execution of single repetitive tasks, with each worker performing only one operation in the overall process. Because of its effect on workers' knowledge, *de-skilling* is a term strongly associated with mass production. And as skill disappeared, so did the unique voice of the craftsman.

The organization was elegantly simple, if not terribly humane. Atop the management hierarchy resided near-omniscient knowledge of products and manufacturing methods. In the case of Ford, product design, process design, marketing strategy, and other critical functions were chiefly the province of one man, Henry. This knowledge was translated into work orders that were executed by an increasingly layered cadre of lieutenants who directed a large but largely unskilled workforce. This style of command-and-control management worked best for single product-lines with few parts and simple processes.

Economies of Scope: Would You Like Fries with That?

Mass production, mass marketing, and mass media have constituted the Holy Trinity of American business for at least a hundred years. The payoffs were so huge that the mindset became an addiction, a drug blinding its users to changes that began to erode the old axioms attaching to economies of scale.

These changes were gradual at first. Even early on, "economies of scope" began to be perceived. General Motors broke Ford's run on the Model-T— an impossibly long product cycle by today's standards—by offering cars that were not black, and even came in different styles to suit different tastes and pocketbooks. Heinz discovered it could make not just, say, mustard, but "57 Varieties" of condiments in the same factory. Consumers began to have a wider range of choice, and they warmed quickly to their new options.

But things got more complicated on the management side. As more products were launched, organizations became increasingly bureaucratic and business functions more isolated from each other. This was de-skilling of a higher order: Design, production, and marketing knowledge began to fractionate, and in some cases, to atrophy.

The real watershed came when offshore producers, finally recovered from the Second World War, began to penetrate U.S. markets. With the oil embargo of the early 1970s, small, fuel-efficient cars began looking highly attractive to people stalled in long gas lines. Companies like Honda, Toyota, and Volkswagen exploded into the North American market like a tsunami. The challenge to U.S. manufacturers was not to offer just trivial feature alternatives, but whole new designs. In a classic reversal, what was suddenly good for America was anything but good for General Motors. The auto industry didn't see these changes coming, and as a result lost enormous market share to offshore competitors.

Overnight, global competition turned mass markets into thousands of micro markets. Nike now makes hundreds of different styles of shoes. THE WALL STREET JOURNAL coined the term *sneakerization* to describe a phenomenon affecting nearly every industry.

Competition is healthy, we'd been told from birth, because it breeds greater choice. But now competition was out of control and old-guard notions of brand allegiance evaporated like mist in the rising-sun onslaught from Japan, Southeast Asia, and Europe. Choice and quality ruled the day, and consumer enthusiasm for the resulting array of new product options forever undermined the foundations of yesterday's mass-market economy.

The relentless search for market niches drove a steep increase in new product introductions, which in turn required an exponential increase in design and process knowledge. There were just two problems. First, mass-production-oriented business processes had been "stove-piped" into noncommunicating bureaucratic business functions. Second, workers, long told to "check your brain at the door," were ill-equipped for the dynamic changes about to wreak havoc on the corporation.

In short, command-and-control management didn't work so well anymore. Necessary knowledge no longer resided at the top. It was as if the organizational core had melted down, and companies that couldn't adjust fast enough— or that were culturally unwilling to shift gears—went belly up as a result.

Who Knows?

This sudden need for more, better, and better-distributed knowledge spawned various attempts at a solution. Three are especially noteworthy.

1. **Concurrent Engineering:** What if separate functions—say design and manufacturing—talked to each other from the outset of a product cycle? This astoundingly obvious idea hadn't yet occurred to anyone because market hegemony and mass production had made it appear unnecessary. If you made only one product, and it had a long life cycle, there was no problem. However, as products proliferated and life cycles accelerated, the need to manage widely distributed knowledge became intense. While concurrent engineering was a step in the right direction, it assumed there was

sufficient knowledge in top-down control functions to specify
detailed commands to thousands of workers producing hundreds
of different products. Big mistake.

2.   **Artificial Intelligence:** Announced with messianic fanfare in the
1980s, this new branch of computer science sought to automate
expertise. If "scientific management" had ramped productivity
through the division of physical labor, why not apply the same
techniques to intellectual labor? However, if industrial automation
is de-skilling, AI is akin to a frontal lobotomy. Instead of distributing knowledge, so-called expert systems made it dependent on
complex and inflexible software. In most cases, these programs
simply didn't work. Knowledge worthy of the name is highly
dynamic. It requires deep understanding, not just rules and algo-
rithms. While machines are lousy at this sort of thing, people are
remarkably adaptive and intelligent. People learn. Real expertise
is changing too fast today to lend itself to automation.

3.   **Total Quality Management:** TQM suggested the unthinkable to
companies intent on automating knowledge: why not look to your
employees? The basic idea was to empower the people who actually did the work. Knowledge resides within practice—a truth
that AI forgot, to its fatal detriment. In companies that adopted
some form of TQM, business practices began to resemble older
notions of craft instead of the brain-numbing repetition of pre-
ordained procedures. People were encouraged to share what they
knew with each other, with other departments and divisions, and
with the company as a whole. This exchange became a rapidly
expanding conversation—a conversation that would soon populate
the corporate intranet.

Understanding, learning, exploration, curiosity, collaboration—qualities that
had been bred out of workers by industrial management—were now being desperately elicited by the All-New, Culturally Revolutionized Organization. Many
spouted the new religion, but secretly tried to hedge old bureaucratic bets.
A handful walked the talk, but it was tough going. A central tenet of TQM was
W. Edwards Deming's dictum: "Drive Out Fear"—a challenge that went to the
heart of the corporation. Conversations among workers were finally seen as
critical to the spread of valuable knowledge—"best practices" in the still-

current jargon. Conversations are where intellectual capital gets generated. But business environments based on command-and-control are usually characterized by intimidation, coercion, and threats of reprisal. In contrast, genuine conversation flourishes only in an atmosphere of free and open exchange.

Enter the Internet

Our whirlwind historical tour has focused on manufacturing because that sector was first to experience these changes. Later, the same forces began to reshape service and information industries. The Internet not only arrived into the context of a newly globalized economy, it has been profoundly shaped by it. Companies installing intranets are seeking to capture and preserve critical knowledge. Individuals coming onto the Internet are seeking the same range of choice that was first offered by imported cars and stereo equipment.

However, most "e-commerce" plays today look a lot like General Motors circa 1969—looking for that next lucrative mass market just when markets have shattered into a million mirror-shard constituencies, many asking for something altogether different from the mindless razzle-dazzle of the tube. Marketeers still drool at the prospect of the Net replicating the top-down broadcast model wherein glitzy "content" is developed at great cost in remote studios and jammed down a one-way pipe into millions of living rooms. TV with a buy button! Wowee!

Today, many large companies offer flashy bread-and-circus entertainments on the Web. These offerings have all the classic earmarks of the mass market come-on: lowest-common-denominator programming developed to package and deliver market segments to mass merchandisers. This is not what most people want, or they would have stuck with television, the Yellow Pages, and 800 numbers. And they don't have to accept it since the Internet came up with the concept of infinite channel-surfing.

The Net represents cheap natural resources (data), cheap transport (the pipe itself), and most important, cheap and efficient access to global know-how. The barriers to entry have fallen so low that a huge number of companies can now compete for a niche—an influx that echoes the entry of Asian and European competitors into U.S. markets. But this is more like an invasion from outer space: Ten thousand saucers just landed and they're merely the advance wave.

Just as GM mistook the Hondas and VWs for a passing fad, most corporations today are totally misreading this invasion from Webspace. Their brand will save them. Right. Their advertising budget will save them. Uh-huh. More bandwidth will save them. Sure. Well, ...something will save them. They're just not too sure what it is yet. But the clock is now ticking in Internet time. Maybe they should get a clue. And quick.

Border Crossings

To most large traditional companies, the notion that workers might actually know what they were doing was a huge insight. (Duh!) But it takes hard work to implement the changes required to elicit knowledge from employees. In most cases, that work is not only incomplete, it hasn't even begun. "Drive out fear"? Dream on.

Knowledge worth having comes from turned-on volitional attention, not from slavishly following someone else's orders. Innovation based on such knowledge is exciting, inflammatory, even "dangerous," because it tends to challenge fixed procedures and inflexible policies. While collaboration has been paid much lip service within corporations, few have attempted it beyond their own boundaries. Ironically, companies that remain "secure" within those boundaries will be cut off from the global marketplace with which they must engage in order to survive and prosper.

And this engagement must be fearless and far-reaching. Workers must become fully empowered and self-directed. Scary. Suppliers must become trusted allies in developing new products and business strategies. Scarier still. Markets must come to have faces and personalities in place of statistical profiles. Flat-out panic!

For many, the new landscape is barely recognizable, online or off. Where business is headed there are no roadmaps yet, and few comforting parallels with the past. The landscape has little to do with mass production, mass merchandising, mass markets, mass media, or mass culture.

Instead, the future business of businesses that have a future will be about subtle differences, not wholesale conformity; about diversity, not homogeneity; about breaking rules, not enforcing them; about pushing the envelope, not punching the clock; about invitation, not protection; about doing it first, not doing it "right"; about making it better, not making it perfect; about telling the truth, not spinning bigger lies; about turning people on, not "packaging" them; and perhaps above all, about building convivial communities and knowledge ecologies, not leveraging demographic sectors.

The New Workplace: Breaking the Silence

*Let us speak, though we show all our faults
and weaknesses—for it is a sign of strength
to be weak, to know it, and out with it...*
Herman Melville

JUST AS TRADITIONAL MEDIA CONDITIONED THE AUDIENCE to be passive consumers—first of commercial messages, then of products— the traditional organization conditioned employees to be obedient executors of bureaucratically disseminated work orders. Both are forms of broadcast: the few dictating the behavior of the many. The broadcast mentality isn't dead by any means. It's just become suicidal.

In contrast, the Internet invites participation. It is genuinely empowering, well beyond the cliché that word has become. And corporate intranets invite participation in the same way. There are strong reciprocal parallels between the open-ended curiosity of the new marketplace and the knowledge requirements of the new organization. The market-oriented Internet and workforce-focused intranet each relies on the other in fundamental and highly complementary ways. Without strong market objectives and connections, there is no viable focus for a company's Internet presence; without a strong intranet, market objectives and connections remain wishful thinking.

The same technology that has opened up a new kind of conversation in the marketplace has done the same within the corporation, or has the potential to do so. But many businesses, especially large ones, still refuse to acknowledge these radical shifts affecting internal workforces and external markets. They don't want to relinquish hierarchic control. They don't want to give up the tremendous economies of scale they enjoyed under the old-school broadcast-advertising alliance. It's what they know. It's how they made their fortunes. However, trying to keep things in the old familiar business-as-usual rut denies the ability of markets to respond to and interact with companies directly— and this is what the Internet has brought to the party.

Why the denial? Could it be that companies are afraid the Internet and intranets will make people smarter? While no company would ever admit to it publicly, this is precisely what many fear. In the "good old days," consumers weren't expected to make suggestions or ask for new features. They were simply supposed to buy the product—any color they wanted as long as it was black. In the same way, workers weren't expected to offer insights or sugges-tions, just to do what they were told.

Networks greatly facilitate the sharing of relevant knowledge within a com-munity joined by like interests. As a result, the lowest common denominator of informed awareness tends to be much higher online than it ever was in the context of broadcast media. Plus, this informed awareness tends to increase much faster. This accelerated learning effect obviously applies to intranets as well—it's where their primary value lies. But a lot can get in the way of this value before it has a chance to evolve and mature.

In 1995, BUSINESSWEEK ran an excellent cover story on intranets, just around the time the buzzword was emerging into general parlance. Several CIOs were quoted as saying they had so-and-so many thousand Web pages behind their firewalls. They were crowing about it. But my take was that this content didn't get created top-down by the organization. Instead, these pages sprang up overnight like a crop of magic mushrooms on a rich mother lode of corporate horseshit.

What does that mean, you ask? Well, look, when all this got started you had thousands of workers with easy access to free Web browsers and a smaller set of folks who had figured out how to set up Web servers whose only cost was download and tinkering time. These people soon figured out that HTML wasn't rocket science, and it was off to the races from there. Suddenly there was nothing to prevent the expression of their own ideas and creativity. Skunk-works wanted to build broader support for their projects, individuals wanted to be noticed for their technical savvy or penetrating wit or business insight.

But then the big-O Organization discovered what was going on, and often as not, brought all this self-motivated fever-pitch development to a grinding halt. Hey, way to go!

To be fair, there were a few high-level execs out there who truly understood the dynamics of how this stuff worked. And by dynamics, I mean more the cultural aspect of networking. For the technology, you could buy a book. Aside from this handful, though, most corporate managers were clueless in the extreme.

And, sadly, most still are. Too many have never spent any serious time online. Then, when they get charged with building a corporate intranet, the first thing they think about is reporting structures and where everybody will sit in some abstract org chart. But dictatorial directives—"All Web pages must be formally approved by the Department of Business Prevention"—throw cold water onto all that magic-mushroom enthusiasm.

The fact is, people at the bottommost tiers of the organization often have far more valuable knowledge than managers and corporate control freaks. If you kill off this enthusiasm, you can easily end up with a large, professional-looking, and very expensive intranet that nobody gives a damn about. The question companies should be asking themselves is: What if we built an intranet and nobody came?

Top management support needs to come in the form of funding, facilitation, and enough brains to get out of the way. It's gotta be more like rock and roll than strait-laced traditional business—and that puts the Suits right over the edge. It's just not possible, they argue, to run a business by letting everybody improvise.

But companies function that way whether anybody wants them to or not. Nobody really runs them; no one writes the score. Corporate management is still largely unaware of what's going on in the marketplace. But their workers know, because they're operating there already. What's going on is the Internet.

Today, market expectations are solidly welded to Net-speed performance. Your software product isn't available for downloading? You don't have secure transaction processing so I can buy it when I need it? Hey, I'm gone! And so is a big chunk of your market share. If your company feeds me a ration of facile hype instead of answering my questions, I'm looking for another supplier.

And the expectation of getting quick, straight answers applies across the board to information of every stripe. It applies to ideas—how to acquire them within the company and from the market, move them around, sort them, slice them, dice them, move them back out into the market as new products, get customer feedback—then iterate, getting better as you go. Make mistakes. Debug on the fly. It's fast, it's furious. It's fun! If you want a rock-and-roll company, which is more important, adhering to procedure or knowing how to dance?

The fervor that produced the first wild-oats crop of intranets surely didn't come from the CIOs who got quoted in BUSINESS WEEK. Workers have had it with repressive management that just gets in the way. Markets have had it with hyperbole-laden corporate rhetoric that's 99 percent hot air. The next huge opportunity for business is to bring workforce and market together. And companies smart enough to realize this start instigating a potent form of internal anarchy.

Unfortunately, such companies are rare exceptions. Most are hanging on for dear life to the one thing they think they can't live without: control. But they only think they're in control. Feeling their real abilities and contributions have gone unappreciated, many employees simply do what they feel like doing anyway, giving as little as possible to the company. They punch the clock and that's it. The relationship is adversarial as hell. If you look into it closely, though, the company has almost invariably set things up this way—by not trusting people to take the initiative, to be engaged, motivated, intelligent, creative, innovative. It's a long, sad story with roots that go back to the early industrial era.

Corporate intranets represent a prime opportunity to turn this scenario around, but only if there's genuine awareness of where the real challenges lie.

Too much of intranet development is focused on whiz-bang technology and not nearly enough on the cultural revolution all this implies and in fact demands.

In healthy intranet environments, work gets coordinated via cooperation and negotiation among colleagues. But these things happen very fast, not in committee meetings. This is why employees need more power in organizations —not to lord it over others, but to make intelligent decisions on the fly and not see them overturned two days later by managers who don't know the territory. Without getting into the politics of it, the biggest complaint of the U.S. armed forces in Vietnam was that the war was being fought from Washington. Again without getting into the politics of it: The United States lost. This is a big clue as to how many intranet initiatives are playing out. Top-down command-and-control management has become dysfunctional and counterproductive.

Imposed infrastructures hinder more than help. Most so-called empower-ment initiatives are embarrassingly paternalistic, to the point of backfiring entirely. Real authority is based on respect for knowledge, and the two are inherently intertwined. Also, both grow bottom-up. When arbitrary "manage-ment" takes over what was initially a handcrafted intranet, the individuals who championed and created it often feel betrayed and disenfranchised. You see the same thing in what happened to craft and the individual voice during the course of the Industrial Revolution. We're making some very old mistakes here.

Take another example much closer to the present. The autonomous PC chal-lenged the hegemony of mainframe computer systems and enabled the develop-ment of quick solutions that could end-run the infamous MIS-bottleneck—the fact that it could take months for computer applications to be created and exe-cuted to deliver needed information. Then IT management discovered the LAN, which delivered another layer of utility. However, instead of leveraging this new resource for the benefit of "users"—even that word is an artifact of the mental-ity—the IT department largely used the LAN to reestablish control over infor-mation access and work environments.

Now, many companies are doing the same thing again with the intranet. You get this rule-book mindset—the corporation's common look and feel, logo placement, legal number of words on each Web page. Whatever. It's all so cramped and constipated and uninviting. Dead. The people who actually built the intranet—created the content that makes it valuable—bail out, looking for another, more open system. And today that's easy to find.

Remember the context for all this. Twenty years ago, or even five, only corporations could provide the kind of resources needed to process even modest volumes of information. The cost of such systems was a significant barrier to entry for new businesses that might become competitors. But today individuals have this kind of power in their rec rooms. And they can get all the Internet they can eat for a few bucks a month. If the company doesn't come through with the kind of information and delivery that turns them on—provides learning, advances careers, and nurtures the unbridled joy of creation—well, hey, they'll just do it elsewhere. Maybe in the garage.

This sort of thing has already been happening for a while now, of course, but there's more on the way, and not just from the usually suspected quarters. To understand what's really happening on the Internet, you have to get down beneath the commercial hype and hoopla, which—though it gets 90 percent of the press—is actually a late arrival. From the beginning, something very different has been brewing online. It has to do with living, with livelihood, with craft, connection, and community. This isn't some form of smarmy New Age mysticism, either. It's tough and gritty and it's just beginning to find its voice, its own direction. But it's also difficult to describe; as the song says, "It's like trying to tell a stranger about rock and roll." And it's next to impossible to understand unless you've experienced it for yourself. You have to live in the Net for a while.

At this level, things are often radically other than they appear. A new kind of logic is emerging, or needs to. I call it gonzo business management—paradox become paradigm. We're not in Kansas anymore, Toto, and we might as well get used to it. There's a huge opportunity here for individuals to keep their day jobs but at the same time to indulge their natural human bent for self-expression.

Companies that try to prevent this sort of creativity within their firewalls need to have their collective heads examined. Conversely, companies that foster and encourage it will win big. The best software, design, music, graphics, writing—elegant, artistic, fantastically interesting and valuable content—are coming out of places where people feel their creativity is valued. Places where inspiration is paramount and posturing means nothing.

Great intranets come from corporate basements, not from boardrooms. How do you know where the next big thing is going to come from? You need great radar today, and that means a wide-awake workforce that's constantly tinkering, exploring, and figuring out new ways to have fun.

The long history of distrust between workers and management didn't start with Karl Marx or the AFL-CIO. It's based more on fallout from the ideas of people like Frederick Taylor and Henry Ford, ideas like "scientific management" and Theory X. Underlying these questionable principles that have done so much to shape the assumptions of business-as-usual is the premise that workers are lazy, unwilling, even stupid. Today, this premise translates into the near-certainty that employees are pilfering company time, collecting a paycheck while hanging out on the Web all day. They probably are. But that's a symptom, not a cause.

The people who built the first intranets put in ridiculously long days. They worked like soldiers rebuilding a bridge. You had to be there to believe it. But many now managing Internet or intranet projects were not there and they don't believe it. It all goes back to fear of losing control. Whatever the motive, the mentality has to go. Right now these fear-driven corporations are spending millions on market research, the whole point of which is to find out who their customers are. They don't know anymore. They've barricaded themselves in their executive suites, and now they've erected firewalls on top of that.

Sure, data security is necessary and needs to be done well. However, many corporations are desperate for firewalls because they don't want the market to see they have nothing worth stealing inside them. That's not security, it's para-noia. You can't identify best practices without sticking your neck out—but if you don't, you risk premature death. You can't invite customers to contribute design ideas by holding them at bay.

And unless your industry is very "mature"—which really means ready for the bone yard—your market isn't wearing pinstripe suits anymore, either. Many companies are currently doing market planning today using straw-man models of the customer that constitute a bad pastiche of Eisenhower-era sitcom out-takes and those throwback Human Resources manuals that haven't been edited in thirty years. Was anybody ever this straight or this stupid? Are they now? If not, what does this say about current approaches to online marketing? In many cases, your workers are your market. Come out of the bunker once in a while, see what they're up to—it could be your future.

But for that to happen, you've to get beyond the firewall. The Internet/intranet dichotomy reinforces the "not invented here" syndrome that has damaged so many companies. Corporations have long understood that they have to tear down the internal walls that prevent necessary cross-functional communication. Now they have to tear down their external walls as well. The survivors will be left standing naked—the stuff of nightmares for many companies. But they'll be left standing naked in the middle of a thriving marketplace. For businesses capable of grasping the ramifications, this is an enormously promising paradox.

In a networked market, the best way for a company to "advertise" will be to provide a public window on its intranet. Instead of putting up slick images of what they'd like people to believe, corporations will open up so people can see what's really going on.

Sometime soon, companies will have to open up significant portions of their intranets—while still protecting their few genuine secrets—in order to create relationships with their markets rather than barriers against them. Otherwise, they're saying in effect: "We know everything we need to know. Why should we look beyond our own borders?" That's just plain wrong, and everybody knows it—especially your workers and your customers.

Companies that are actually communicating with online markets have flung the doors wide open. They're constantly searching for solid information they can share with customers and prospects via Web and FTP sites, e-mail lists, phone calls, whatever it takes. They're not half as concerned with protecting their data as with how much information they can give away. That's how they stay in touch, stay competitive, keep market attention from drifting to competitors. Such companies are creating a new kind of corporate identity, based not on the repetitive advertising needed to create "brand awareness," but on substantive, personalized communications.

The question is whether, as a company, you can afford to have more than an advertising-jingle persona. Can you put yourself out there: say what you think in your own voice, present who you really are, show what you really care about? Do you have any genuine passion to share? Can you deal with such honesty? Such exposure? Human beings are often magnificent in this regard, while companies, frankly, tend to suck. For most large corporations, even considering these questions—and they're being forced to do so by both Internet and intranet—is about as exciting as the offer of an experimental brain transplant.

But the future looks dismal only to companies that are spooked by the prospect of coming in out of the cold. Those at highest risk aren't wonderful places to be working in at any level today. Their future could be very bright if they'd just decide to stop being prisons with nasty wardens. If they choose not to stop, I don't have much pity. Companies that are harming themselves out of ignorance can, with a little humility and a lot of hard work, begin to learn and change. I've seen it happen, and it's an impressive thing. On the other hand, companies that are harming the people who work for them out of cowardice, greed, and willful stupidity richly deserve whatever fate may have in store.

Giant companies tend to look only over the tops of the trees at other giants they consider worthy competitors. Few bother to look down at their feet. If they did, many would see their foundations being nibbled away by competitors many times smaller, yet fiercely committed to do battle for even a tiny slice of this new territory. Some little garage operation can only take away, say, .001 percent of market share from one of these monster companies. However, a hundred thousand garage operations can take it all—and given the new business dynamics the Internet brings to bear, this can happen overnight. The Net will cause radical discontinuities, catastrophic breaks in the already crumbling façade of business-as-usual.

Companies currently have a lot of motivation to get serious. And to get really serious, they first have to get a sense of humor and relax—yet another pretzel-logic paradox. They need to relax to break the obsessive-compulsive control habit. They need to understand that employees already know how to do the work far better than the company could ever hope to dictate. Corporate intranets could unleash the potential energy of the corporation, but to nourish and grow that potential, companies have to relinquish their addiction to management. Zen master Suzuki Roshi once said, "To control your cow, give it a bigger pasture."

At some point you've got to break down and trust people both inside and outside "your" organization—and the Web is responsible for those quotation marks. It is radically blurring the boundaries of what's inside and outside, yours and theirs. The only way companies can sound authentic to new online markets is to empower employees who actually have the knowledge to disseminate it on their behalf. And from here on out, that's always going to mean a two-way street between workplace and marketplace.

The New Marketplace: Word Gets Around

In the late eighteenth century, the British philosopher Jeremy Bentham imagined a little nightmare he called a "panopticon"—a prison in which the inmates could be seen at all times, but couldn't see their jailers. A few hundred years later, mass media inverted this scenario. The imprisoning TV eye now sees nothing, yet we all watch it for clues to our cultural identity. But what would happen if each of these isolated prison cells were somehow wired to all the rest so the inmates could observe their overseers? Not only see them, but also speculate about their motives, and compare notes on their behavior and intentions? It's already happened. That's what the Internet does. Suddenly the overseer is like an insect mounted on a pin for all to view.

While corporations are still only marginally aware of what's being said about them online, all but the totally out-of-it are uncomfortably aware these conversations are taking place, and that the control they had in the days of broadcast has evaporated. We're not just watching the ads these days, we're publicly deconstructing them. In this context, intranets look like salvation to many companies, their protective firewalls a form of corporate encryption designed to insulate against a scary new kind of market: unpredictable, unmanageable, unwilling to be manipulated.

At one point the Cluetrain Manifesto says: "Markets do not want to talk to flacks and hucksters. They want to participate in the conversations going on behind the corporate firewall. De-cloaking, getting personal: We are those markets. We want to talk to you."

De-cloaking even more: I wrote that last bit. Personally. The Internet has radically changed the nature of the marketplace. I believe this. But how do I presume to know it? Certainly not through market-research reports, most of which aren't worth the paper they're written on. I know it because the Internet has changed me and the thousands of people I talk to every week. Maybe the best way to explain this is to tell my own story—talk about who I am and how I got here. Am I representative of the online market? The point is that there is no "online market" in some general abstract sense. More than any market that's ever existed, the Internet is a collection of unique individuals. I'm one of them.

I bought my first computer in 1981. It had a 300-baud modem that I used to connect to The Source, the first commercial online service. For those who may not know, *baud* is a technical term meaning "extremely slow." Nonetheless, I used this machine to talk to people I'd never met. We'd hook up and say things like: "Hey, who are you? What's happening over there? And by the way, where is over there?" The personal computer seemed to me the all-purpose machine. You could draw with it, write on it, save thoughts and recall them later, recombine them—you could even make music. I was a carpenter and a cabinetmaker and into tools in a big way. Here was a machine that communicated with others of its type, and behind each one was another person, another mind jamming, improvising, conveying ideas, feelings, and experiences I'd never before had a way to tap into. I'd never encountered a tool this powerful.

Through a weird combination of fortuitous accidents, I ended up in Tokyo several years later working in an artificial intelligence project for the Japanese government. What the project needed—and what I had to offer—was a fairly good grasp of the English language. What I lacked was any formal training in computer science. Nothing had prepared me for the stratospheric high-tech world I suddenly found myself immersed in. I knew next to nothing about machine intelligence, but I was fascinated by its core concept of "knowledge engineering." The challenge was to model how people understand things, represent ideas, and communicate them to others. In this case, the "other" was a computer. I could relate to the enormity of the problem. I was groping around in the dark myself, struggling with new concepts, and learning as I went. I was flying by the seat of my pants.

One day, I met with a researcher in a coffee shop. Language was a problem, but he spoke more English than I did Japanese. I had just been to the bookstore and was lugging a stack of books on highly advanced computer-science topics. It was all Greek to me, but I figured something might rub off. Suddenly the guy asks me, "Who gives you permission to read those books?"

I was stunned. Bowled over. Did his puzzlement reflect some sort of cultural difference? I didn't think so. It struck me that this fellow was just being more honest and direct than an American might be. He was articulating what many people in today's world seem to assume: that official authorization is required to learn new things. I thought about this deeply, and I'm thinking about it still.

Who gives us permission to explore our world? The question implies that the world in fact belongs to someone else. Who gives us permission to communicate what we've experienced, what we believe, what we've discovered of that world for ourselves? The question betokens a history of voice suppressed, of whole cultures that have come to believe only power is sanctioned to speak. Because the ability to speak *does* involve power. It entails ownership and the control conferred by ownership. As the saying has it: "Money talks, bullshit walks."

Right then and there, in that chance encounter in some random Tokyo coffee shop, I gave myself blanket permission: to be curious, to learn, to speak, to write. But it's a long road from permission to practice, and there's plenty of negative reinforcement in between. Freedom of expression may be called out loftily in the U.S. Constitution, but even after two centuries of democracy, it's still a far cry from second nature.

Communication is a powerful tool. And like any other powerful tool, it has been pressed into the service of business-as-usual. A few years after my stint in Japan, I ended up back in the United States, hired by an AI software outfit to be their director of corporate communications. Cool, I thought. That sounded important. I had no idea what it meant. Only later did I discover I'd become their PR guy. Bummer.

I was pretty naïve back then, but I quickly figured out that public relations was perceived by the press—the people I was supposed to be talking to— as little more than thinly disguised hucksterism. I tried playing the high-tech huckster role precisely once and came away from the experience feeling dirty, phony. I couldn't bring myself to do it again, which was a big problem. It was my job. And I needed the money. Stop me if any of this sounds familiar.

The "key messages" of any AI software company back then involved head-bangingly abstruse concepts like "heuristics," "backward chaining," and "non-monotonic logic." Very deep. And *very* boring. I barely understood this jargon myself. How was I supposed to get on the phone with some total stranger and enthuse about The Product? The truth was, I didn't give a damn about the product. What I cared about was knowledge, how people acquired and used it, how organizations suddenly seemed to need a lot more of it, and why. What I cared about was how technology applied—or didn't—to the world of business and the actual people who worked there.

So instead of pitching the product, I started talking to journalists about stuff like that. I figured I'd just pretend to be working until I got fired for goofing off. But something amazing happened. As soon as I stopped strategizing how to "get ink" for the company that was paying my salary, as soon as I stopped seeing journalists as a source of free advertising for my employer, I started having genuine conversations with genuinely interesting people.

I'd call up editors and reporters without a thought in my head—no agenda, no objective—and we'd talk. We talked about manufacturing and how it evolved, about shop rats and managers, command and control. We talked about language and literature, about literacy. We talked about software too of course—what it could and couldn't do. We talked about the foibles of the industry itself, laughed about empty buzzwords and pompous posturing, swapped war stories about trade shows and writing on deadline. We talked about our own work. But these conversations weren't work. They were interesting and engaging. They were exciting. They were fun. I couldn't wait to get back to work on Monday morning.

Then something even more amazing happened. The company started "getting ink." Lots of it. And not in the lowly trade rags it had been used to, but in places like THE NEW YORK TIMES and THE WALL STREET JOURNAL and BUSINESS WEEK. One day the CEO called the VP of Marketing into my office.

"What has Chris been doing for you lately?" the CEO asked him.

"I'm glad you brought that up," said the marketing veep. "In the whole time he's been here, he hasn't done a single thing I've asked him to."

"Well…," said the CEO looking down at his shoes—here it comes, I thought, this is what it feels like to get sacked—"whatever it is he's doing, leave him alone. From now on, he reports to me."

That's how I discovered PR doesn't work and that markets are conversations.

That's also how I started ghostwriting for the CEO. One afternoon I was banging out an article, and I wrote a paragraph that stopped me cold. It stopped me because something new and very different had just showed up on the screen: my own voice. It's hard to explain, but the paragraph I'd just written resonated with something that had been sleeping all my life, something potent, something deep. I realized I could say things I cared about, and I could say them in a way no one else could. I stopped ghosting and started writing my own stuff.

But it was hard to write the sort of thing that gave me that same feeling. Where could I publish it? I would try to sneak some of myself into the articles I wrote for journals and magazines, but I usually had to disguise what I really wanted to say.

In 1995, I ended up in IBM's Internet division. A ranking PR guy from corporate headquarters ran into me one day and said he'd heard I had a lot of contacts in the financial press. He suggested we get together for lunch and talk about it. I took this as a good sign, maybe an opening to do what I liked best. But when we met several weeks later he said something like, "All those journalists you know? Never talk to them again."

He said I should refer all such conversations to him instead. That way, he said, the company's messaging would be consistent. Or words to that effect. But I knew they wouldn't be real conversations—they would be "key message" pitches, and I wasn't about to subject people I knew and liked to that sort of targeting. I kept my contacts to myself.

I was devastated. It was bad enough that I'd been explicitly forbidden to speak with journalists, many of whom had become good friends, but where was I going to *write*? If I published anything, I'd get busted for not asking permission—there was that word again—and if I wrote sleazy PR for IBM, I'd have to kill myself to blot out the karmic stain.

And then it came to me: I could write on the World Wide Web! At that juncture, IBM's Internet division was so clueless I figured most of the top brass had only vaguely heard of it. One senior guy thought Yahoo was a kind of browser—no lie—and this was after the Yahoo IPO had made headlines in every major newspaper worldwide. Oh well, at least their PR was consistent.

I liked this idea. A lot. I'd be invisible on the Web, outside the control of any company. I'd be free at last to speak in my own voice without begging anyone's permission. I decided to create a Web-cum-e-mail newsletter. I wanted a catchy title, so I called it ENTROPY GRADIENT REVERSALS, EGR for short. In the beginning, I thought it would be a perfect vehicle to deliver my profound pundit-grade insights about the Internet and show everyone how smart I was. That didn't last long. I ended the very first issue like this:

> ❏ From time to time we offer to share our list of subscribers with
> door-to-door aromatherapy salespersons and ritual ax-murderers.
> If you would prefer that your data not be used in this way, please
> check the box.

Whoa! What a response that brought! Everyone was laughing. People subscribed in droves. I was ecstatic. I wondered whether IBM would have given me permission to publish such material. Probably not—on the off-chance of offending the aromatherapy and ritual ax-murderer market segments.

I started wondering what other sorts of noncorporate things I could write. What if I broke all the rules? You know, the unwritten rules everyone learns by telepathy at birth: be pleasant, be brief, don't speak down to your reader, don't use big words, don't use obscenity, don't make yourself the center of attention. First and foremost, do that all-important market research. Find out what your audience wants to hear about. Ask their permission.

Wait a second...hadn't I been through all this? I had, and I'd had enough.
I decided to go against the grain with a vengeance. I told readers they were
clueless hosers. I interviewed an imaginary horse—at exhausting length. I used
vocabulary so obscure that people needed unabridged dictionaries to figure out
what I was saying. I developed an alter-ego named RageBoy®, a seriously mal-
adjusted mental case and towering egomaniac with an advanced case of
Tourette's syndrome. And my readers loved it.

Well...the ones who stayed loved it. Many went screaming for the nearest
exit. RageBoy at full throttle is not everyone's cup of tea, to be sure. But
the ones who stayed are an interesting lot. Some are programmers, teachers,
artists, writers, full-time parents. Others have titles like Director of Public
Relations, VP Marketing, Chief Information Officer, CEO. And the companies
they come from read like the Fortune 500 list. The readership is not, as you
might suspect, drawn from some dangerously misanthropic idiot fringe. The
audience is regular people, mostly business people. And as the THX ads say:
The audience is listening.

Forget my gonzo experimentation with RageBoy. That's just one microscopic
example of what's happening online. The real point is that the Internet has
made it possible for genuine human voices to be heard again, however different
they may be from the cautious, insipid pabulum of mainstream broadcast media.
Why has the Internet grown so rapidly? Why did it catch so many businesses off
guard? The audience is listening because people are attracted to precisely the
difference the Net provides: the sound of human beings talking with one
another as human beings—the sound of a million conversations whose primary
purpose, for once, is not to sell us something.

How do these conversations get started? How do people with common
interests find each other? How does anyone find anything online? The simple
answer is the theme of this book: Word gets around. And on the Net, word
gets around fast.

For every entry in the encyclopedia, there is now a Web site. For any
idea you can imagine—and some you can't—there are thousands of articles
and images electronically swirling around the globe. But that's not the real
story. That's not the big news. The word that's going around, the word that's

finally getting out, is something much larger, far more fundamental. The word that's passing like a spark from keyboard to screen, from heart to mind, is the permission we're giving ourselves and each other: to be human and to speak as humans.

Consciously or not, millions of us are using the Internet to pass along this unconditional permission to millions of others. When enough people do that, something viral happens. It's not hypothetical, it's happening—when we say what we think, when we feel what we say, when we listen for the music of authentic presence. We are constantly searching each other out, linking, talking, shaking things up. Consciously or not, by the very nature of the permission we give each other, we are working to bring down business-as-usual.

News of this ever-spreading word is what you're reading here. And it's a little schizophrenic, I have to admit. In one sense, the news is good. It's great! It's the joyous noise of people reveling in a newfound freedom, laughing, jeering, cheering, irrepressible.

From another perspective, the news is not good at all. Everybody's miserable. Everybody's had about enough. People are sick to death of being valued only as potential buyers, as monetary grist for some modern-day satanic mill. They're sick of working for organizations that treat them as if they didn't exist, then attempt to sell them the very stuff they themselves produced. Why is a medium that holds such promise—to connect, to inspire, to awaken, to enlist, to change—being used by companies as a conduit for the kind of tired lies that have characterized fifty years of television? Business has made a ventriloquist's trick of the humanity we take for granted. The sham is ludicrous. The corporation pretends to speak, but its voice is that of a third-rate actor in a fourth-rate play, uttering lines no one believes in a manner no one respects.

Oh, well. That's OK. We'll get by. We've got each other.

I have to laugh as I write that. The Internet audience is a strange crew, to be sure. But we're not talking about some Woodstock lovefest here. We don't all need to drop acid and get naked. We don't need to pledge our undying troth to each other, or to the Revolution, or to the bloody Cluetrain Manifesto for that matter. And neither does business.

All we need to do is what most of us who've discovered this medium are already doing: using it to connect with each other, not as representatives of corporations or market segments, but simply as who we are.

From hopelessly romantic meditations on favorite cats, to screeds so funny you'll blow coffee out your nose, to collective code for alternative operating systems: We're all expressing ourselves in a new way online—a way that was never possible before, never before permitted. And make no mistake, speech once freed is a powerful drug. Get used to it; it ain't going back in the box. What does this mean for electronic commerce? Take a wild guess. We're not those neatly predictable consumers business remembers from yesterday. We got a taste of something else, and we like it. We'll make it ourselves, and defend it with a ferocity that might surprise most businesses. If you're a business, believe us: it's a surprise you'd just as soon skip. We're in the market for lots of things, but the market we see ourselves in is more like that ancient marketplace. Tell us some good stories and capture our interest. Don't talk to us as if you've forgotten how to speak. Don't make us feel small. Remind us to be larger. Get a little of that human touch.

people of earth . . .

*Your choice is simple. Join us and live in peace
or pursue your present course and face obliteration.
We shall be waiting for your answer.
The decision rests with you.*

The Day the Earth Stood Still (1951)

To find anything that isn't overtly complicit with the Great Technology Sitcom, you have to dig down to the underbelly of the Web. You have to get past the sites with commercial pretensions that are slicing and dicing you, counting the legs and dividing by four, bringing in the sheep. You are being incorporated into their demographic surveys. And, predictably, the lowest common denominator is getting all the juice. You are being packaged for advertisers by some of the hippest hucksters on the planet.

Dig deeper. Down to the sites that never entertained the hope of Buck One. They owe nobody anything. Not advertisers, not VC producers, not you. Put your ear to those tracks and listen to what's coming like a freight train. What you'll hear is the sound of passion unhinged, people who have had it up to here with white-bread culture, hooking up to form the biggest goddam garage band the world has ever seen.

What are these underbelly sites about? What's a rock concert about? How about creation, exploring a visceral and shared collective memory we've been brainwashed into believing never existed?

Conspiracy theory, my ass. Schools and teachers, the motor vehicle bureau, the IRS, the military, the line at the bank, the television set, the newspapers at the checkout stand, the news on your radio, the billboards along the highway, and now a hundred thousand cold-comfort Web sites. All are tuned to your brain at the deepest level and you have lined up for the coolest, latest-model implant. The carrier wave has been tuned at huge cost to deliver a single message: You are not free, you desire nothing but the products we produce, you have no world but the world we give you.

If you're OK with this, then eat it up. There's a bulimic's dream-feast of killer kontent on the way. But if it already makes you want to puke, get angry. Write it, code it, paint it, play it—rattle the cage however you can. Stay hungry. Stay free. And believe it: Win, lose, or draw, we're here to stay. Armed only with imagination, we're gonna rip the fucking lid off.

There's your market.

Prospectus

OK, THE SCARY PART IS OVER NOW. YOU CAN COME OUT. It's safe.

In fact, the news gets better from here on out. And the first bit of news is that this isn't about us and them. It's about us. *Them* don't exist. Not really. Corporations are legal fictions, willing suspensions of disbelief. Pry the roof off any company and what do you find inside? The Cracker Jack prize is ourselves, just ordinary people. We come in all flavors: funny, cantankerous, neurotic, compassionate, avaricious, generous, scheming, lackadaisical, brilliant, and a million other things. It's true that the higher up the food chain you go, the more likely you are to encounter the arrogant and self-deluded, but even top management types are mostly harmless when you get to know them. Given lots of love, some even make good pets.

Inside companies, outside companies, there are only people. All of us work for organizations of some sort, or we're peddling something. All of us pay the mortgage or the rent. We all buy shoes and books and food and time online, plus the occasional Beanie Baby for the kid. More important, all of us are finding our voices once again. Learning how to talk to one another. Slowly recovering from a near-fatal brush with zombification after watching NIGHT OF THE LIVING SPONSOR reruns all our lives.

Inside, outside, there's a conversation going on today that wasn't happening at all five years ago and hasn't been very much in evidence since the Industrial Revolution began. Now, spanning the planet via Internet and World Wide Web, this conversation is so vast, so multifaceted, that trying to figure what it's about is futile. It's about a billion years of pent-up hopes and fears and dreams coded in serpentine double helixes, the collective flashback déjà vu of our strange perplexing species. Something ancient, elemental, sacred, something very very funny that's broken loose in the pipes and wires of the twenty-first century.

There are millions of threads in this conversation, but at the beginning and end of each one is a human being. That this world is digital or electronic is not the point. What matters most is that it exists in narrative space. The story has come unbound. The world of commerce became precipitously permeable while it wasn't looking and sprang a leak from a quarter least expected. The dangers of democracy pale before the danger of uncontained life. Life with the wraps off. Life run wild.

Why do companies find this prospect terrifying? How, for instance, does the above description differ from the basic operation of a virgin forest? When he said, "in wildness is the preservation of the world," I bet Thoreau wasn't just thinking about old-growth trees. He also wrote a little ditty called ON CIVIL DISOBEDIENCE. There is a connection.

But don't look at us. For the defenseless position all you companies now find yourselves in, you can thank the creators of the Internet, the U.S. Department of Defense. What a paradox. What a total hoot!

And you might as well hoot as cry about it. It's not the end of the world. It's the beginning of a new one. What's emerging is our story in the most fundamental sense, the human mythos weaving a vision of whatever it wants to become. There is no known deterrent. Take a deep breath, baby. Roll with it.

While this may sound spooky and mystical and terribly uncorporate, it isn't meant to put you Fortune 500 types off. When you get right down to it, human beings are spooky and mystical and terribly uncorporate, and corporations—if you'd only let yourselves admit it—consist entirely of human beings. Sort of neat how that works out. So the bottom line is: You can play in the Internet headspace as well as anyone.

There are just three conditions: 1) you have to let your people play for you, since there's really nobody else at home; 2) you have to play, not something more serious and goal-oriented; and 3) related to the previous, you have to have at least some tenuous notion of what "headspace" might mean. It's not in the dictionary. But you can ask around. Get the general hang of the thing. If you figure it out, we'll think you're cool and consume mass quantities of all your wonderful products.

See how easy life can be when you loosen up a little?

You laugh, we laugh with you.

Either way, we live.

CHAPTER 2

The Longing

DAVID WEINBERGER

What Is the Web For?

WE KNOW TELEPHONES ARE FOR TALKING WITH PEOPLE, televisions are for watching programs, and highways are for driving. So what's the Web for?

We don't know. Yet we put it on magazine covers, found businesses stoking it, spend billions on an infrastructure for it. We *want* it to be important with a desperation that can frighten us when we look at it coldly.

Who is this we? It's not just the webheads and full-time aficionados. It's the management teams who don't understand it but sense an opportunity. It's the uncles and aunts who pepper you with questions about all this Web stuff. It's the seven-year-old who takes it for granted that when she speaks the entire world can choose to hear her. Our culture's pulse is pounding with the Web.

This fervid desire for the Web bespeaks a longing so intense that it can only be understood as spiritual. A longing indicates that something is missing in our lives. What is missing is the sound of the human voice.

The spiritual lure of the Web is the promise of the return of voice.

Being Managed

THE LONGING FOR THE WEB OCCURS IN THE MIDST
of a profoundly managed age.

We believe, in fact, that to be a business is to be managed. A business manages its resources, including its finances, physical plant, and people in basically
the same way: Quantifiable factors are determined, predicted, processed,
assessed.

But our management view extends far beyond business. We manage our
households, our children, our wildlife, our ecological environment. And that
which is unmanaged strikes us as bad: weeds, riots, cancer.

The idea that we can manage our world is uniquely twentieth-century and
chiefly American. And there are tremendous advantages to believing one lives
in a managed world:

- **Risk avoidance.** Nothing unexpected happens if you're
  managing your world.

- **Smoothness.** Everything works in a managed environment
  simply because broken things are an embarrassment.

- **Fairness.** In earlier times, life was unfair. Now you're guaranteed
  your three-score and ten and if something "goes wrong," the
  managed system will compensate you, even if you have to sue
  the bastards.

- **Discretionary attention.** If you were out in the wild, your attention would be drawn to every creaking twig and night howl. But
  now that the risks have been mitigated, things work right, and
  you can manage your time so you have not just leisure time but
  also discretionary attention: You can decide what interests you.
  Why, you can even have hobbies.

Of course, none of these benefits are delivered perfectly. There are still risks,
there are still injustices, there are still "outages." But these are exceptions. And
when they occur, we feel cheated, as if our contract has been violated.

It wasn't always thus. For millennia, we assumed that being in control was
the exception and living in a wildly risk-filled world was the norm:

> *As flies to wanton boys, are we to the gods.*
> *They kill us for their sport.*
>
> **King Lear**

Today these awful words sound like one of those quaint, primitive ideas we've outgrown.

The belief in the managed environment is a denial of the brute "facticity" of our lives. The truth is that businesses cannot be managed. They can be *run*, but they exist in a world that is so far beyond the control of the executives and the shareholders that "managing" a business is a form of magical belief that gets punctured the first time a competitor drastically lowers prices, a large trading partner's economy falters, a key supplier's factory burns down, your lead developer gets a better offer, your CFO becomes felonious, or an angry consumer wins an unfair lawsuit.

As flies to wanton boys are companies to their markets. They pull off a company's wings for sport.

How to Hate Your Job

A MANAGED ENVIRONMENT REQUIRES BEHAVIOR FROM US that we accept as inevitable although, of course, it is really mandatory only because it is mandated. We call it "professionalism."

Professionalism goes far beyond acting according to a canon of ethics. Professionals dress like other professionals (one eccentricity per person is permitted—a garish tie, perhaps, or a funky necklace), decorate their cubicles with nothing more disturbing than a Dilbert (formerly Far Side) cartoon, sit up straight at committee meetings, tell carefully calibrated jokes, don't undermine the authority of (that is, show they're smarter than) their superiors, make idle chatter only about a narrow range of "safe" topics, don't swear, don't mention God, make absolutely no reference to being sexual (exceptions made for male executives after the hot new hire has left the room), and successfully "manage" their home life so that it never intrudes unexpectedly into their business life.

Most of us don't mind doing this. In fact, we actually sort of enjoy it. It's like playing grownup. And having extremist political banners hung in cubicles or having to listen to someone talk about his spiritual commitments or sex life would simply be distracting. Disturbing, actually.

And yet...we feel resentment. Find someone who likes being managed, who feels fully at home in his or her professional self. Our longing for the Web is rooted in the deep resentment we feel towards being managed.

However much we long for the Web is how much we hate our job.

Our Voice

JUST ABOUT ALL THE CONCESSIONS WE MAKE TO WORK in a well-run, non-disturbing, secure, predictably successful, managed environment have to do with giving up our voice.

Nothing is more intimately a part of who we are than our voice. It expresses what we think and feel. It is an amalgam of the voluntary and involuntary. It gives style and shape to content. It subtends the most public and the most private. It is what we withhold at the moments of greatest significance.

Our voice is our strongest, most direct expression of who we are. Our voice is expressed in our words, our tone, our body language, our visible enthusiasms.

Our business voice—in a managed environment—is virtually the same as everyone else's. For example, we learn to write memos in The Standard Style and to participate in committee meetings in The Appropriate Fashion. (Of course, we are also finely attuned to minute differences in expression and can often tell memos apart the way birdwatchers spot the differences between a lark sparrow and a song sparrow.)

In fifty years, our corporate lives will seem no different than those of the 1950s. Whether we are Ward Cleavers or Dilberts, we all reported to work in look-alike rooms, wearing uniforms, speaking civilly, playing our parts at committee meetings. The fact that earth tones and Rockports have replaced gray flannel and wingtips isn't going to separate us from our crewcut fathers.

Managed businesses have taken our voices. We want to struggle against this. We wear a snarky expression behind our boss's back, place ironic distance between our company and ourselves, and we don't want to think we have become our parents. But we have. And we've done so willingly.

Management is a powerful force, part of a larger life-scheme that promises us health, peace, prosperity, calm, and no surprises in every aspect of our lives, from health to wealth to good weather and moderately heated coffee from McDonald's. We are all victims of this assault on voice, the attempt to get us to shut up and listen to the narrowest range of ideas imaginable.

It is only the force of our regret at having lived in this bargain that explains the power of our longing for the Web.

The Longing

WE DON'T KNOW WHAT THE WEB IS FOR BUT WE'VE ADOPTED it faster than any technology since fire.

There are many ways to look at what's drawing us to the Web: access to information, connection to other people, entrance to communities, the ability to broadcast ideas. None of these are wrong perspectives. But they all come back to the promise of voice and thus of authentic self.

At the first InternetWorld conference, the vendors were falling over one another offering software and services that would let you "create your own home page in five minutes." Microsoft, IBM, and a hundred smaller shops were all hawking the same goods. You could sit in a booth and create your own home page faster than you can get your portrait sketched on a San Francisco sidewalk.

While the create-a-home-page problem proved too easy to solve to support a software industry, there was something canny about the commercial focus on the creation of home pages. Since you could just as adequately view the Web as a huge reference library, why did home pages seize our imaginations? Because a home page is a place in which we can express who we are and let the world in. Meager though it may be, a home page is a way of having a voice.

The Web's promise of a voice has now gone far beyond that. The Web is viral. It infects everything it touches —and, because it is an airborne virus, it infects some things it doesn't. The Web has become the new corporate infrastructure, in the form of intranets, turning massive corporate hierarchical systems into collections of many small pieces loosely joining themselves unpredictably.

The voice that the Web gives us is not the ability to post pictures of our cat and our guesses at how the next episode of THE X-FILES will end. It is the granting of a place in which we can be who we are (and even who we aren't, if that's the voice we've chosen).

It is a public place. That is crucial. Having a voice doesn't mean being able to sing in the shower. It means presenting oneself to others. The Web provides a place like we've never seen before.

We may still have to behave properly in committee meetings, but increasingly the real work of the corporation is getting done by quirky individuals who meet on the Web, net the two-hour committee meeting down to two lines (one of which is obscene and the other

Make a Bonfire of Your Reputations

When I was asked to make this address I wondered what I had to say to you boys who are graduating. And I think I have one thing to say. If you wish to be useful, never take a course that will silence you. Refuse to learn anything that implies collusion, whether it be a clerkship or a curacy, a legal fee or a post in a university. Retain the power of speech no matter what other power you may lose. If you can take this course, and in so far as you take it, you will bless this country. In so far as you depart from this course you become dampers, mutes, and hooded executioners.

As a practical matter a mere failure to speak out upon occasions where no opinion is asked or expected of you, and when the utterance of an uncalled-for suspicion is odious, will often hold you to a concurrence in palpable iniquity. Try to raise a voice that will be heard from here to Albany and watch what comes forward to shut off the sound. It is not a German sergeant, nor a Russian officer of the precinct. It is a note from a friend of your father's offering you a place in his office. This is your warning from the secret police. Why, if any of you young gentleman have a mind to make himself heard a mile off, you must make a bonfire of your reputations and a close enemy of most men who would wish you well.

I have seen ten years of young men who rush out into the world with their messages, and when they find how deaf the world is, they think they must save their strength and wait. They believe that after a while they will be able to get up on some little eminence from which they can make themselves heard. "In a few years," reasons one of them, "I shall have gained a standing, and then I will use my powers for good." Next year comes and with it a strange discovery. The man has lost his horizon of thought. His ambition has evaporated; he has nothing to say. I give you this one rule of conduct. Do what you will, but speak out always. Be shunned, be hated, be ridiculed, be scared, be in doubt, but don't be gagged. The time of trial is always. Now is the appointed time.

—JOHN JAY CHAPMAN
Commencement address
to the graduating class
Hobart College, 1900

wickedly funny), and then—in a language and rhythm unique to them—move ahead faster than the speed of management.

The memo is dead. Long live e-mail. The corporate newsletter is dead. Long live racks of 'zines from individuals who do not speak for the corporation. Bland, safe relationships with customers are dead. Long live customer-support reps who are willing to get as pissed off at their own company as the angry customer is.

We are so desperate to have our voices back that we are willing to leap into the void. We embrace the Web not knowing what it is, but hoping that it will burn the org chart—if not the organization—down to the ground. Released from the gray-flannel handcuffs, we say anything, curse like sailors, rhyme like bad poets, flame against our own values, just for the pure delight of having a voice.

And when the thrill of hearing ourselves speak again wears off, we will begin to build a new world.

That is what the Web is for.

CHAPTER

Talk Is Cheap

RICK LEVINE

The voice emerges literally from the body
as a representation of our inner world.
It carries our experience from the past,
our hopes and fears for the future,
and the emotional resonance of the moment.
If it carries none of these, it must be
a masked voice, and having muted the voice,
anyone listening knows intuitively
we are not all there.

David Whyte, *The Heart Aroused*

Voices from Pots

I'M A POTTER'S KID. WHEN I WAS GROWING UP, WE ALWAYS had red-brown carpet and rugs to help hide the terra-cotta dust we tracked home from Dad's shop. I have fond memories of watching Italian potters with doorway-wide shoulders spin clay into forms larger than myself, effortlessly raising planters, lamps, bowls, and jars from undistinguished lumps of mud, one after the other, parading dozens of seemingly identical forms across the studio floor. Whenever I see a large thrown shape, I remember the first time I tried to throw twenty-five pounds of clay, thinking I would start with less than the sixty or seventy pounds I saw growing like graceful mushrooms on my dad's wheels. I landed flat on my back, shoulder blades bruised, smelling twenty years of clay dust on the wood floor beneath my head, as the misshapen lump of clay took advantage of my first indecision, knocking me from the wheel and covering the studio wall in red mud. You can't learn to throw large forms without losing lots of them in the process.

Pots are made by people. Large ones especially remind me of that human authorship. Smaller things—mugs, cups, pitchers—touch me as well. They're fitted to a potter's hands, reflecting their measure. I can gauge the size of the artisans' hands, the length of their fingers, from lips, spouts, and pulled handles. There's so much more life invested in a thrown piece than in anonymous cast or stamped ware. A medium such as clay, elevated and transformed by human shaping, bears witness to the life that molded it into something more than plain stuff.

When experienced potters describe their craft, they often talk about seeing the form they're creating in their mind's eye, applying force to make the spinning clay match its virtual, internal archetype. There's an incredible amount of practice, failure, and learning that has to take place before we develop the courage and surety to trust such an internal, private muse, to ignore the contrary opinions of others and do what we know will succeed.

Despite too many years spent behind keyboards and display screens building software, creating Web sites, and generally using technology more than is good for me, I'm still a potter's kid. I consider myself an artist and a craftsman, and bring a craftsman's attitudes to my work and life. One perspective that seems to surface with some regularity is a deeply instilled obligation to do new work, create stuff people have never seen before. It's a peculiar approach to life, picked up mostly by osmosis at some early age from my parents and relatives. In execution, it's a standard requiring constant exploration and reinvention, but also a certain studied ignorance of what's considered right and proper. There's a bit of irrationality in believing that if I follow my own intuition and, at some level, don't pay attention to what other people think, I'll create unique works that will surprise and delight. Artists have a stubborn faith in their ability to create newness from next to nothing. This faith shapes their work, enables them to establish themselves as individuals, fingerprinting their way through their medium.

What's this got to do with business? With organizations? Lots. Most of the creative people and knowledge workers organizations depend on, those whose sense of self-worth is centered in the pride they take in the work of their heads and hands, will have an immediate "been there, done that" reaction to this description of artistic identity. From the electronic pressroom gang, to the MIS boiler-room toilers, to the hackers building an insurance-entitlement management app to increase next year's sales—all have some of the attitude of the craftsman. People in high tech take pride in their work. They are individuals who see the details of the things they produce in the light of the trials and triumphs they experience while creating products. In the courage of creation, they find a place to hang their individuality. Programmers and techno types appreciate elegant, spare code and the occasional well-turned architectural hack. My accountant friends get off on clever spreadsheet macros, and on being able to slant this quarter's results to shade meaning within the arcane constraints of the law. Even managers leave their telltale fingerprints on their jobs—the well-coached team rising to unexpected heights, or the business relationship blossoming into a long-term sales annuity. Some people apply a craftsmanlike care to their work, and their voices are heard, remarked upon, and recognized as uniquely theirs.

The Web is no different. Every Web page we see has a person behind it. Sometimes their individual decisions are eroded and digested by being passed through a corporate colon of editors, gatekeepers, and other factota, but there are clear signposts to individual care and concern on much of today's Web. While all print and broadcast media have at least some

Claude Levi-Strauss, the French anthropologist, discusses *bricolage* as the opportunism of those who work with their hands, creating stuff out of whatever is lying about. The Web is group bricolage. Individuals build it without working from a master plan. They take pieces that work—stealing gifs, formats, links—and create new pages. This makes the Web unpredictable, creative, and always the result of human hands.

—David Weinberger

indirect personal authorship, there's a key difference on the Web. The percentage of "raw" content published, direct from a creator's fingers to our eyes, is much higher than in traditional media. The Web's low cost of entry to publishers, both small and large, and the amount of unfiltered chat/newsgroup/e-mail text finding its way into search engines guarantees our daily browsing experience has a very strong flavor of individual authorship. Inevitably, our heightened awareness of distinct, individual voices engenders the urge to talk back, to engage, to converse. The software and mechanisms developed helter-skelter for the Net cater to these urges. Chat, free e-mail, automatic home pages—all reinforce our feeling that not only is it easy to enter into discourse with others, but also that we're by-god *entitled* to wade into the conversational stream. Heaven help you if you get in my way, or try to stifle my voice.

The good or bad news, depending on your perspective, is that it's hard to fake your end of one of these conversations. Ever been on the phone with a friend or coworker while sitting in front of a computer and trying to read or respond to e-mail when your wire addiction gets the better of you? I'm very good at multitasking, and can fool many folks some of the time, but I get caught more often than I'll admit. (By my wife, for instance. Every time.) We can tell when someone is engaged, listening, responding honestly, and with his or her full faculties. We're wired to interpret subtle clues telling us whether a person is all there, if we're the center of their attention, if we're being heard. No matter how starved for detail our communication channel, our brains manage to get a gestalt reading on the other party's presence.

Rick may believe he's good at multitasking, but I don't believe it. Humans can't multitask—we can't pay attention to two things simultaneously.

You can multitask? Fine. Then read a book and write one at the same time. No, multitasking is really just rapid attention-switching. And that'd be a useful skill, except it takes us a second or two to engage in the new situation we've graced with our focus. So, the sum total of attention is actually decreased as we multitask. Slicing your attention, in other words, is less like slicing potatoes than like slicing plums: You always lose some of the juice.

—David Weinberger

In the same way we distinguish personal attention from inattention, we can tell the difference between a commercial pitch and words that come when someone's life animates their message. Try snipping paragraphs of text from press releases and a few pieces of printed person-to-person e-mail. Shuffle the paper slips. Hand the pile to your office-mate, your spouse, or your next-door neighbor. Can they sort them? Of course they can, in short order. People channel from their hearts directly to their words. That's voice. It comes of focus, attention, caring, connection, and honesty of purpose. It is not commercially motivated, isn't talk with a vested interest. Talk is cheap. The value of our voices is beyond mere words. The human voice reaches directly into our beings and touches our spirits.

Voice is how we can tell the difference between people, committees, and bots. An e-mail written by one person bears the tool marks of their thought processes. E-mail might be employee-to-employee, customer-to-customer, or employee-to-customer, but in each case it's person-to-person. Voice, or its lack, is how we tell what's worth reading and what's not. Much of what passes for communication from companies to customers is washed and diluted so many times by the successive editing and tuning done by each company gatekeeper that the live-person hints are lost.

Authenticity, honesty, and personal voice underlie much of what's successful on the Web. Its egalitarian nature is engendering a renaissance in personal publishing. We of genus Homo are wired to respond to each other's noise and commotion, to the rich, multi-modal deluge of data each of us broadcasts as we wade through life. The Web gives us an opportunity to escape from the bounds imposed by broadcast media's one-to-many notions of publishing. Nascent Web publishing efforts have their genesis in a burning need to say something, but their ultimate success comes from people wanting to listen, needing to hear each other's voices, and answering in kind.

Wired Conversations

THE MESSAGE HERE ISN'T NEW AND ISN'T PARTICULARLY complex. Our elevator pitch is a pretty short one:

> People talk to each other. In open, straightforward conversations. Inside and outside organizations. The inside and outside conversations are connecting. We have no choice but to participate in them.

If there's any newness, it's in how the Net and the Web change the balance of the conversational equation. Technology is putting a sharper, more urgent point on the importance of conversation. Conversations are moving faster, touching more people, and bridging greater distances than we're used to. Let's take a tour of the various conversational modalities the Net offers and how they carry our voices.

E-mail

ELECTRONIC MAIL IS THE WEDGE CRACKING THE ROCK of corporate communication. I write a message, label it with yourname@ wherever_you_are.com, click the "send" button, and you've got my mail. Most corporate electronic defenses pass it right through. They might screen out applets, viruses, and other denizens of the internetworked dark, but words slip through like wraiths. Thoughts. Ideas. Kudos. Complaints. Jokes. We exchange the mundane, day-to-day electronic utterances greasing our business down its intended path. We also trade other missives, possibly words my management would rather I didn't speak, or didn't hear. But the flow can't be stopped, not without choking off the lifeblood of most businesses. The inexpedient comes with the expedient, and we have no choice but to work with it. The basic operating rule of e-mail is that anyone can send mail to almost anyone else—all they need is an address.

E-mail is a more immediate medium than paper. My expectation of the response time to many messages I send is today, not tomorrow or a week from now. This urgency means I'm more apt to write quickly and conversationally when I respond to a message. A lot of the spontaneity in e-mail messages comes from writers breaking through their natural caution and reserve, rushing the writing process, giving themselves permission to be blunt, honest, and sincere in response to a query. It's not just a question of knowing how to type, but of giving myself permission to truly converse: to "out" myself in a conversational medium that is informal, honest, yet open to myriad misinterpretations if I choose words and phrases carelessly. Despite this scary thought, most people don't find person-to-person e-mail daunting. The ease and directness of e-mail is forging new connections—new conversations—throughout virtually every business. Type-click-deliver.

Mailing Lists

MAILING LISTS COME IN TWO BASIC FLAVORS, ONE-WAY and two-way. One-way lists let me send to a large number of people at once, but recipients can't respond to the entire list the message was sent to. There's no opportunity for conversation, other than between you, the recipient, and me, the list owner. These lists can often have the character of a mass mailing, like a Christmas card list. If the mailing is from someone you don't know who's trying to sell you something, convince you of something, or lure you to a particular Web site, it's called *spam*.

There's a fascinating subgenus of the one-way list called a webzine or an e-zine, as in electronic magazine. These are periodic bouts of creative journalism sent to willing subscribers, with audiences ranging from dozens to hundreds of thousands. To their devotees, they have all the interest and attraction of their well-funded offline counterparts. However, they're often more focused, more idiosyncratic, and less plastic. They're usually created in someone's garage or bedroom office as a labor of love given a pulpit by the incredibly low entry cost of Internet publishing. They may start out like the UTNE READER or MOTHER JONES magazine, but the Web relieves them of the need to raise capital, rent a press, and pay for postage. Many 'zines have a strong conversational tone. They mine the incoming stream of responses to take the temperature of their constituencies, and relay tasty bits back to their audiences. The conversation this engenders often feels like publishing with a more immediate feedback loop.

Two-way lists are even more interesting from a conversational viewpoint. They let recipients respond to messages, and everyone else on the list sees their responses. But it takes time to wade through all the traffic on a busy list, sifting value from chaff, knowledge from data. As individual mailing lists grow from small, focused forums, they can easily turn into large, unwieldy free-for-alls. The commitment required to understand the content and context of a list before you post to it is part of the conversational ante this aspect of the Net requires. Just like voice conversations, these asynchronous exchanges reveal if you're all there, focused, and paying attention.

In a moderated two-way list, all mail to the list is screened by someone doing gatekeeper duty. The moderator's role ranges from that of a friendly guide to being an editor with absolute control over every message sent to the list. Moderators often end up having a great deal of influence on the tenor and substance of list conversations.

Conversations on two-way lists look just like conversations in personal e-mail, except the odds of having several people responding independently to the same piece of mail go way up. The conversation may branch, spawn side discussions, and loop back on itself as each new person throws in her or his two cents.

Newsgroups

NEWSGROUPS ARE SIMILAR TO MAILING LISTS, EXCEPT messages collect on special computers on the Internet configured as "news servers." (A business can also have news servers for internal company discussions, not available on the pubic Internet.) I can point a newsreader at the server to check in when I want to, rather than seeing all the messages accumulate in my e-mail in-box, willy-nilly. Newsgroups can be either moderated or unmoderated, just like mailing lists. Newsgroups also record the conversational thread structure of their messages, unlike e-mail, so you can see who is talking to whom and why.

The information space encompassed by publicly available newsgroups, called Usenet, is enormous. Every month, millions of conversations across the globe are enabled by newsgroups. Where e-mail conversations are often between people who know one another, Usenet exchanges are often between strangers. It's a medium that encourages discourse, and can create a kind of community among its participants.

All of these channels for conversation—e-mail, mailing lists, newsgroups—begin to look more alike as you use them. At some point, you start paying more attention to the messages and conversations, and less to the differences in software and tools employed by the various electronic delivery channels.

For our purposes, the biggest difference between electronic and paper mail is the ease with which a single message can be distributed to a vast audience, and then serve as a seed for conversation. I can forward your mail to my friends. To lots of my friends. To lots of people I've never met before, but who might be very interested in what you have to say about your company's management, policies, or practices. I can do this on a scale far exceeding paper distribution. And I can do it before lunch.

Here's a real-life example of how a conversation ignores the boundaries between companies and their customers. It starts with a posting to a newsgroup for Saturn car owners, rec.autos.makers.saturn, an entirely reasonable question about how much service should cost and how much control a car owner has over what gets done to his car. This newsgroup isn't owned or in any way managed by Saturn. It's just plain folks, talking about their cars.[1]

> Subject:
> Am I Getting F-'ed By My Saturn Dealer???
> Date: 1999/06/29
> Author: Ross <rossxxx@xxx.com>
>
> I would like to hear you'alls opinions and experiences. I got a 99 SC2 with now 17k. My owners manual says at 9k and 15k all it needs is an oil change.
>
> At 9k I got charged $50 and they tacked on a 9k Service card that says: Oil Change, Top Off Fluids, check and adjust tires, check all lights in & out, inspect brakes, inspect throttle linkage, lube door & hood hinges.
>
> At 15k I got charged $113 and they tacked on a 15k Service card that says: Oil Change, Top Off Fluids, check and adjust tires, check all lights in & out, inspect brakes, inspect throttle linkage, lube door & hood hinges, inspect cooling system, clean engine, test onboard computer, report tire wear.
>
> Now I recognize that doing that kind of work requires some payment, BUT all I was expecting them to do (and me to pay for) was only an Oil Change. What gives?? At what point of ownership do I get what service Saturn requires as opposed to what my Saturn Dealer requires, and why is there a difference?
>
> Comments are most appreciated especially from the Techs!!
>
> Thanks...............Ross

---

[1] I've shortened this sequence from its original length of about 20 messages, and edited individual messages to save us some reading time. I've also disguised people's names and e-mail addresses, to protect their privacy.

Someone chimes in about how much similar service costs them—customers of two vendors comparing notes, from the comfort of their home or office:

Subject: Re:
Am I Getting F-'ed By My Saturn Dealer???
Date: 1999/06/29
Author: ranger_xxx <ranger_xxx@xxx.xxx.com>

Where are you taking your car for service? Those prices sound high. At the store I go, the 9k and 15k are just oil changes and the charge is less than $27. I'd shop around other nearby Saturn stores and see what they charge.

Then, some tips for getting along with your car dealer—reference information and advice, a testimonial in support of Saturn dealers:

Subject: Re:
Am I Getting F-'ed By My Saturn Dealer???
Date: 1999/06/29
Author: Eric <eric@xx.xx.xxxx.com>

You do have the right to bring your car in at any of these intervals and request that ONLY an oil change be done. If that's what you requested and that's not what they did then I would have a discussion with the service manager and just tell him that you requested only an oil change, you know they recommended more work, but didn't want that, and see what they say. Every service manager I've talked to in about 8 different shops has always been very helpful and appreciates it when these things are brought to their attention.

Eric
94 SL2 HCS

So far, so good, but here's a pointer to another car dealer's service price list. If I were a slightly shady service manager, I'd worry, as my customers have gone beyond anecdotal comparisons to posting real prices. Now it's harder to lie with a straight face. And exposure isn't limited to the dozen or so people who posted to this discussion, but extends to thousands who might search for "saturn customer service" and find this thread:

Subject: Re:
Am I Getting F-'ed By My Saturn Dealer???
Date: 1999/06/29
Author: Al xxx<axxx@xxx.net>

Ross,
Here is the URL for the Retailers in Columbus, Ohio and what THEY
recommend. It has additions to what the Owner's Handbook says
but they typically let me choose to do or not to do the EXTRA items.

Hope this helps!

Al

Some comic relief from the peanut gallery...

Subject: Re:
Am I Getting F-'ed By My Saturn Dealer???
Date: 1999/06/29
Author: newxxx <jasonxxx@xxx.com>

YOU WERE HAD!!!!
SUE SUE SUE

More disgruntlement. No one from the dealership being flamed answered this post. Some readers will assume this omission validates the indictment of Saturn's service, and the company will lose business. Further, since Saturn corporate minions didn't wade into the discussion at this point, Saturn may have lost some prospective customers.

Subject: Re:
Am I Getting F-'ed By My Saturn Dealer???
Date: 1999/06/30
Author: Leexx <leexx@xxx.net>

Midtown Saturn in Toronto does the same thing. $177 for the
60,000 KM service. When I pointed out to the service rep that the
Saturn manual didn't call for coolant change etc. they had to check
with service manager only to agree to not change coolant (as if car
owner's words are not to be trusted). When I picked up the car,
instead of charging me for the usual $30 oil change, they simply
subtracted the cost of coolant change ($90) from $177. And they
also charged me half an hour labor for inspecting vibrating brakes
and noisy steering which was supposed to be covered under
warranty.

The end result was a $145 oil change. And the car was washed, but
not clean as there were many dusty missed spots. To add insult to
injury, they won't drive me to pick up the car because I arranged
the pick up only with the driver, not the service rep. So add another
$25 cab ride.

Even though Saturn makes good cars for the price and sales reps
are nice, some dealer's service dept needs to catch up to the rest of
the corporation.

I'll NEVER GO TO TORONTO MIDTOWN SATURN again and encourage everyone else to boycott them.

I'm buying another new car in 18 months (wife and I each get new
cars every 3 years). This bad experience will certainly be a factor.
It's too bad other nice Saturn dealers will lose out because of this
one dealer.

Now, a different sort of animal. This is a response from a Saturn employee,
a technician, explaining how the game is played. This is an after-hours posting
from a real person, and may or may not have been sanctioned by Saturn. This
represents the firewall bleed-through at the heart of the CLUETRAIN MANIFESTO:
honest comments from employees, unfiltered, going directly to customers.

Subject: Re:
Am I Getting F-'ed By My Saturn Dealer???
Date: 1999/06/30
Author: Bluexxx <bluexxx@xxx.com>

>At what point of ownership do I get what
>service Saturn requires as opposed to what
>my Saturn Dealer requires, and why is
>there a difference?

>Comments are most appreciated especially
>from the Techs!!

Its late for me after a 12 hour Saturn day but I will give a brief
answer. If you want a real detailed answer feel free to ask more
questions.

Saturn Corp. issues its "recommended" services and intervals. For
warranty purposes these are all you need to follow as far as maintenance goes. Saturn Dealers, on the other hand, are free to amend
them in pretty much what ever way they want, intervals and price
included.

I have worked at 2 different dealers. My current one is much smaller
and the prices are much cheaper. There is a $15 difference in labor
rate and the dealers are only 1.5 hrs apart. The items included in
each service are somewhat different but not as drastic as yours. I
have seen dealers include an alignment with every 12k. I have seen

cooling services done at 24k. Lots of variations. I don't personally like it because when you get cars from other dealers they are out of sync.

I asked Saturn Corp. why they don't mandate a stricter policy towards service intervals. They said its illegal for them to force a dealer to do that.

I hope this helps a little. Feel free to ask any question you have. I personally agree that what happened to you sucks. Our dealer the 9k and 15k are oil changes at about $22.

And, in this case, our Net fairy godmother has the last word. Truth, justice, and a testimonial to the power of the Net. Right.

Subject: Re:
Am I Getting F-'ed By My Saturn Dealer???—UPDATE
Date: 1999/07/14
Author: Ross <ross@xxx.com>

My Saturn Dealer agreed I was getting F'ed. AND "in an effort to earn your trust we extend a free 18k service." So I took them up on it and spoke to the guy in charge (since he didn't respond to my e-mail). He said that the Dealer is "rethinking" its 15k service. I guess between these posts and my response to Saturn's 15k service inquiry, they got the message. So, I guess I'm almost even and at 21k will ask for a simple oil change.

Ross

This conversation wasn't simply a business correspondence. It was among lots of people, ordinary folks. These people are writing in their own voices because they want to talk, to help, to contribute. If it's not altruism, it's something close to it—with maybe an occasional touch of revenge. We listen to their voices to decide whom to trust, and we can come to some pretty accurate conclusions about who's on the mark and who's full of hot air.

There's even overt comic relief and some entertainment. We listen carefully to what wasn't said, and who didn't participate in the conversation. The Saturn dealers are conspicuous by their absence. Their silence speaks as loudly as their words might have, had they joined in. The Saturn mechanic was speaking for his company in a new way: honestly, openly, probably without his boss's explicit sanction. He gave away secrets, took a risk, was humanized—and he greatly served the interests of Saturn. He and others like him are changing the way

Saturn supports its customers. And Saturn corporate might not even know it's happening.

This puts a completely different spin on "talk is cheap." The mechanic's e-mail didn't cost Saturn a nickel. He wrote it on his own time. Companies need to harness this sort of caring and let its viral enthusiasm be communicated in employees' own voices. Pay a little, get a lot. Talk is cheap.

Technology is making these conversational needles lots easier to find in the Internet haystack. There are search services anyone can use to find this stuff. General purpose newsgroup searchers like deja.com keep conversations like these online for varying lengths of time. Some for a very long time. The Internet has a wonderfully retentive memory, and we're constantly working to make it easier to retrieve little-used trivia from its magnetic depths. Don't bet against customers' ability to type your organization's equivalent of "lousy saturn customer service" into their favorite search engine and see your latest ugly truth displayed on their computer screen. And they'll chime in and tell their story. Your story.

Chat

CHAT GETS A BAD RAP. THE WEB CANARD SAYS ALL CHAT sessions degenerate into conversations about sex within five minutes. It ain't so. Because it is immediate—taking place in real time—chat can enable conversation that feels more genuine, more substantial, and more human than any other Net channel.

The aspect of chat that still amazes me is that it can compress distance, enabling globe-spanning conversation in a visceral, obvious way. E-mail can connect people over the same distances, but it doesn't trigger the sense of wonder I experience when I see words appearing on my screen typed live by someone half a world away. Here's a snippet of chat conversation between two friends. The chat session was set up following an international artists workshop held in Tblisi, Georgia, and ran for several months. Belt is in London, Annie in Tblisi.

Belt:—*Just Entered The Room*—

Belt: **hello is anyone there?**

Belt: **its Belt here...waiting for a full gas bottle so I can get warm...**

Annie:—*Just Entered The Room*—

Belt: drinking tea, wondering how to make a living...(london transport has just gone up again! £3.80 for a travel card now...don't know how people manage it!)

Annie: 50 tetri in Tbilisi hey?

Belt: crazy,...what do you think of the euro? and custard pies!!

Annie: I guess theres no difference in some way...is it cold in London?

Belt: well its not so cold, just dark...how about there?
are you still snowed in?

Annie: whats that about...over here its snowstorms deep snow, ploughs the airports shutdown

Belt: will you be able to get back?

Annie: spending time face up to snow flakes lying in bed of white sparkly comfort...making snow angels...have you ever made a snow angel?...tis all new to me

Belt: I remember making snow angels on dartmoor, face up to the sky lying on the cold ground...

Annie: I hope to fly Thurs get in Friday...but apparently theres another storm on the way...its incredibly beautiful, esp. on the industrial landscape of the lake (which looks like the sea...huge)

Belt: it was like trying to fall off the earth....sounds incredible...immense landscapes. I haven't seen the sky since Mirzaani

http://www.devoid.demon.co.uk/generator/chat4.htm

One definition of community is a group of people who care about each other more than they have to. This isn't a business exchange, even remotely. It is conversation, the verbal glue binding people separated by geography into a community. Chat's important to us corporate types because it's a medium where it's almost impossible to operate within the old rules. Because chat is a "live" medium, there's little leeway for faking a voice, for a sophist approximation of a person. You can adopt a new persona, but you're going to need to button it all the way up and live it, or we'll be able to tell there's someone else underneath. Chat is CB radio on steroids. It's immediate and unwashed. If you can't type and think at the same time, you're in deep weeds. We can't broadcast, can't message, can't spew corporate pabulum in a chat environment. If business could successfully integrate chat into its marketing universe, companies would be on their way to shaking off some of the mass-media shackles separating them from customers.

One of the more interesting uses of a chat service has been to provide live customer support for Web sites without resorting to expensive telephone call centers. liveperson.com is selling a service using a pop-up chat client connected to a 24/7 call center to provide live, on-the-spot support to Web customers. Each support rep can field up to four simultaneous chat sessions. Customers get an immediate, interactive support person to answer their questions.

> Constantin: **Hi there, rick. What can I help you with today?**
>
> rick: **Hi Constantin. Could you tell me if your service can work with query string tokens instead of cookies?**
>
> Constantin: **Can I just clarify what you are asking...are you looking to track users using querystrings**
>
> rick: **No, I need a solution that works even if a person refuses to accept cookies, or if their firewall/browser rejects cookies.**
>
> Constantin: **or are you referring to being a user and receiving a cookie when you begin to chat**
>
> rick: **Yep.**
>
> Constantin: **If you hold on one second I can find out for you...**
>
> rick: **Great. Thanks.**
>
> Constantin: **I'm sorry but it is not possible right now to offer the service without sending cookies**
>
> rick: **Ah. Is it planned?**
>
> Constantin: **Just a moment please...**
>
> Constantin: **Not at the present time**
>
> rick: **Ok. Thanks for taking the time.**
>
> Constantin: **Thanks for visiting**

There's none of the hit-or-miss multiday waiting we get with e-mail support, no phone cost to the customer or vendor. Commerce sites have reportedly been experiencing dramatically increased sales from the high-touch attention they can give their customers.

Web Pages

THE WEB LETS US LOOK into other people's lives in an intimate way. It enables us to see people as they are, close up. Have you ever been browsing for information, read an interesting page, followed the author's name link, and tripped through his personal pages, read his badly written poetry, looked at pictures of his dog, cat, family, friends, and trip to the Bahamas?

For instance, while browsing slashdot.org, I read a comment from Chris Worth, followed a breadcrumb trail to www.chrisworth.com, and was captivated by his personal Web phantasmagoria—including a cogent comparison of users of Microsoft productivity tools to frogs in pots of heating water, and

C hat forces us to be human and fully present. Except for those of us who start out as gorillas...

HaloMyBaby: **Welcome, Dr. Patterson and Koko, we're so happy you're here!**

DrPPatrsn: **You're welcome!**

HaloMyBaby: **Is Koko aware that she's chatting with thousands of people now?**

LiveKOKO: **Good here.**

DrPPatrsn: **Koko is aware.**

HaloMyBaby: **I'll start taking questions from the audience now, our first question is: MlnyKitty asks Koko are you going to have a baby in the future?**

LiveKOKO: **Pink**

DrPPatrsn: **We've had earlier discussion about colors today**

LiveKOKO: **Listen, Koko loves eat**

HaloMyBaby: **Me too!**

DrPPatrsn: **What about a baby? She's thinking...**

LiveKOKO: **Unattention**

DrPPatrsn: **She covered her face with her hands...which means it's not happening, basically, or it hasn't happened yet.**

LiveKOKO: **I don't see it.**

http://www.envirolink.org/kokotranscript.html

No, I didn't make it up! Worried about 800-pound media gorillas dominating the Net?

a scary little piece about his view of helicopters. Another time, I searched for how-to information about a software program I was installing, found an article written by Glenn Fleishman, clicked on his byline link to www.glenns.org and was engrossed in his story of how he fought Hodgkin's disease and won.

Early on in Sun's exploration of the Web, our best site efforts came from small teams with broad skills who didn't have too many constraints placed on what they were supposed to create. They had to work quickly and invent solutions to problems nobody had ever faced before. Perhaps because of the variety of skills the teams had to integrate and the short time lines, customers said they thought the sites looked like the result of people having fun and moving quickly. We never had the time to "corporatize" them.

I do this frequently. In my mind's eye, I watch myself clicking off my intended path, wondering what the draw is, why am I allowing myself to be diverted from my goal. It's because I enjoy listening to people. They give me windows into their lives, providing substance as a foil to the superficial factual gloss of their day jobs. I'm seduced into spending time staring at evidence of their humanity, despite my better judgment against such a "waste of time." And then I do it again. And again.

The fact that Web pages are conversations hasn't sunk in because they look like publications. But they are conversations: expressions of individual voice looking for response. The Web pages we revisit often have feedback mechanisms and change over time in response to that feedback. Further, they must change visibly, or people won't come back. We expect change, reaction, reflection of our comments and feedback. This is not just true with respect to personal Web pages. There's a very strong desire for corporate Web pages to have a human feel—to speak to us in some genuine way. This desire cries out for communication that's less formal, less professional, less anonymous, and more for the people reading than for the company doing the writing.

Hart Scientific, Inc. (www.hartscientific.com) posted a convenient comparison of conversational versus traditional writing on their Web site. They have two versions of their Y2K compliance page. You can tell them apart:

> Noncompliance issues could arise if Hart Scientific manufactured products are combined with other manufacturer's products. Hart cannot test all possible system configurations in which Hart manufactured products could be incorporated. Our products currently test as being compliant and will continue to operate correctly after January 1, 2000. However, customers must test integrated systems to see if components work with Hart Scientific manufactured products. Hart makes no representation or warranty concerning non-Hart manufactured products.

And...

> If you're using our equipment with someone else's gear, who the hell knows what's going to happen. We sure don't, so how can we promise you something specific, or even vague for that matter? We can't, so we won't. However, we love our customers and like always we'll do whatever is reasonable to solve whatever problems come up, if there are any.

We seem to know, intuitively, when something spoken, written, or recorded is sincere and honest—when it comes from another person's heart, rather than being a synthesis of corporatespeak filtered by myriad iterations of editing, trimming, and targeting. There's an inherent pomposity in much of what passes for corporate communication today. Missing are the voice, humor, and simple sense of worth and honesty that characterize person-to-person conversation.

We survived Y2K, but that's not the biggest challenge we face. The need for honest speech, to ordinary people, hasn't gone away. Web-savvy consumers are ignoring online brochures. An organization, as presented via the Web, must have a human voice, must stand for something, mean something, want to meet people, and show they're trying to understand those people.

Millions and Millions Served

How do you scale one-to-one networking to reach
thousands and millions of like-minded netizens?
Sure, I could do it with a phalanx of smart people
reaching out and touching electronically, but then my
fledgling company's burn rate would increase faster
than I could raise capital. Where is the balance
there? Seems like any mass-produced message
(even tailored for a given "market") will be
disingenuous to the savvy.

—Jody Lentz, e-mail to cluetrain.com

IS HAVING CONVERSATIONS WITH LOTS OF CONSTITUENTS
really practical? Yes. Our conversations are already reaching more customers
than we know. People have other means of hearing conversations besides talking to us directly. They can "eavesdrop" on conversations we have with others
by reading other people's e-mail posted to the Web, or by reading posts in
newsgroups. The volume of conversation about us we don't participate in
directly is almost always greater than the volume we are personally involved
in. We respond not only to the honesty and integrity of our conversations on
the Net but also to those indicators of integrity in other people's conversations.
Our choice was never to be in all the conversations, but to be honest and open
in those we do engage in.

Companies will survive employees telling their truths, their stories in a
business context, without instituting draconian controls on their ability to
speak out when and to whom they please. We listen to individuals differently
than we do to organizational speech. When a company publishes PR, it's trying
to give us a complete message about who they are and what they do. We have
to decide to trust or distrust the company based on a single statement. Well-
written PR leaves us with few avenues for corroboration and second opinions.
It's meant to be self-contained.

On the other hand, when I converse with people inside a company, I hear
stories from individuals. They're all grains of sand, their combined voices richer

and more diverse than the univocal speech of corporate mouthpieces. We add up all the anecdotes we hear from individuals. We have to trust our own averaging, our own summing of stories, our own divining of truth. With more people, more stories in the mix, it's harder for one negative story to sway me. This speaks to the need to have many people in an organization talking to customers. A single "corporate story" is a fiction in a world of free conversation. Corporate stories, like corporate cultures, are informed by individuals over time through many contacts, conversations, and opportunities to tell stories.

Stories play a large part in the success of organizations. With stories, we teach, pass along knowledge of our craft to colleagues, and create a sense of shared mission. Will coordinating what a large number of people have to say be a problem? Yes and no. The problem is not in the effort required to coordinate voices, but in the attitude that assumes speech demands coordination and control. A culture of story-telling, one encouraging the collection and sharing of knowledge in conversation, may

There is no longer a single source for "The Truth." You can no longer download the corporate PR campaign as reality and go from there. You are now hearing many voices, many "truths," and will have to pick and choose and integrate. A company's fear is that a lone voice with an axe to grind will make up a "truth" as plausible as anything the Marketing department has come up with, but harmful to the company, and it'll be adopted as the "truth" irrespective of the facts.

The reality is that when malicious propaganda happens (and it will happen), the truth will out. That is the glorious thing about the markets of opinion; no opinion stands unchallenged. This is the real fear many corporations have of the markets of opinion, that their white propaganda— they are the Galahads of the industry, pure and good and perfect—will also not survive the challenges of the market of opinions.

They're right, they won't. Cope. The corporations will be revealed to be made up of fallible human beings.

Just like us.

—Brian Hurt, *e-mail to cluetrain.com*

There's legal risk in freer communication between employees and customers. Companies accustomed to issuing pronouncements from a single, tightly controlled department find this conversational shift somewhat terrifying. The intersection of the webbed world with business-as-usual leaves much legal ground uncharted. While we're waiting to see how our laws will evolve to meet these realities, it might be prudent for companies to consider which is more damaging: silence, or talking to customers in many individual voices. Is the legal risk posed by unfettered speech always a valid excuse for not speaking?

need encouragement and example-setting, but it will certainly fail in the face of attempted restraint.

When we were building Sun's first Intel-based workstation, the 386i, we used mock magazine reviews of the product as a way to test ideas for the design of the computer and the software. As the design progressed, we settled on one "review" as an example of a magazine article that might appear when the product shipped. The ersatz review was a hit with team members: It became a decision yardstick for months of subsequent design and implementation questions. People also started giving copies of the review to customers and using it as a conversation starter with friends and colleagues in other companies. The review wasn't a product pitch—it required a person to deliver it, explain it, and fill in lots of details. It wasn't a data sheet, but a foil for stories and conversations. Its value was not in creating some kind of official spin, but in enabling the reliable transfer of knowledge and new ideas.

A critical aspect of success with large numbers of customers lies in listening to them. It's not enough for employees to talk to customers. There must be a way for the fruits of employee conversations to trickle back into an organization's plans. When Sun started to address the problem of providing technical support to the Java developer community, we made a glaring error. We assumed our answers to technical questions were more valuable than answers from sources outside our group, than answers from our customers.

Sun's first launch of the Java Developer Connection Web site was an unabashed effort to package a fairly sleazy business proposition: selling per-

incident support for a poorly supported and less than adequately
documented software product.
We were doing a lousy job of
helping Java developers. A bright
marketing wag had the idea to sell
people answers to their questions
for one hundred bucks a pop.
When a licensing engineer who
dealt with customers day in and
day out posed the question, "Why
should they feel good about paying us for answers that should be
in the docs, or for consulting on
problems caused by the instability
of our products?", the marketing
folks decided to use a bit of
sugarcoating. For $495 a year, a
customer could purchase a "subscription" including five questions, called "support incidents,"
and a package including technical
newsletters and other goodies.
Unfortunately, the marketing
team focused on providing the
hundred-dollar support answers
and didn't spend much time setting up the information-publishing
pipeline for the sugarcoating. We
had fewer than two hundred pay-

Here's a definition of that pesky and borderline elitist phrase, *knowledge worker*: A knowledge worker is someone whose job entails having really interesting conversations at work.

The characteristics of conversations map to the conditions for genuine knowledge generation and sharing: They're unpre-dictable interactions among people speak-ing in their own voice about something they're interested in. The conversants implicitly acknowledge they don't have all the answers (or else the conversation is really a lecture) and risk being wrong in front of someone else. And conversations overcome the class structure of business, suspending the org chart, at least for a little while.

If Knowledge Management aims at making organizations smarter, and that in turn means increasing the quality of conversa-tions, can we please avoid the temptation to create something called Conversation Management?

—David Weinberger

ing customers for the service. Almost all used up their magic answers in less
than a month, then started clamoring for the "real" value, the (grossly under-
staffed) information subscription that was to be their pipeline to successful use
of Java. To add insult to injury, our own cross-divisional inefficiencies cost us
$110 for each question answered. You do the math.

Time for phase two. We shut down the site, and relaunched a free service with a few critical new features. The staffing problem hadn't gotten better, so we brainstormed ideas for getting the Java community to help us solve their problems. We now have a free site with question-and-answer forums where developers answer each other directly. We added a tap into Sun's Java software bug database and provided a means for developers to add their own notes and work-arounds to our bug information, as well as vote for the bugs they wanted us to fix soonest. A reverse pipeline into the company sent the bug votes to our engineers to help with prioritization. The site hit one million registered members in two years, a far cry from the two hundred in six months that the initial, traditional support efforts yielded. Moreover, the site became a nexus for conversations about our products and services, and for conversations about other people's solutions to our problems.

Symantec took a similarly creative approach when they first launched their Café product, a suite of programming tools for Java developers. They had one person virtually living in the public support newsgroups. He responded to questions, fielded tech support requests, and generally got himself known as a very straight shooter about Symantec's products. He was only one person, but he was almost single-handedly responsible for the developer community's positive take on Symantec. He wasn't there to promote, but strictly to assist. He gave honest answers to hard questions, acknowledged product shortcomings, and painted an honest, open picture of the product's strengths and weaknesses. The developer community's collective opinion of Symantec soared.

Another anecdote from the public relations history of Sun's Java team paints an anti-example. In the first year and a half that Sun's Java group existed, members of the engineering team spoke directly with customers and the press. Java grew from a glimmer, a possibility, to a platform with thousands of curious, turned-on early adopters. There was a general perception that Sun's Java team listened, answered questions, and was actively engaged with the community of Java developers.

After about eighteen months, the workload grew to such a point that we started shutting down our channels to the outside world. PR and marketing took over much of our contact with the outside world, and we put our heads down to deal with the increasing demands on the engineering team. The reaction from our developers was stated in these precise words many times over: "you disappeared." As we went underground, the perception of the Java group in the marketplace changed from "a small team of great engineers producing neat stuff" to "a hype engine to push Sun's stock." In projects that allowed engineers time to come back online more often, customers cut the engineers far more slack in their attempts to get things right than did the customers of their more close-mouthed brethren.

In such scenarios, of course, engaging in trivial conversations can chew up valuable time—though it's a tough call to know what's truly trivial and what deepens credibility in the conversational space. Sometimes responding to a joke with a one-line e-mail laugh can do wonders. We still need to answer all the mail, but we can do things to eliminate some of the more repetitious communication. Generally, people inclined to find answers themselves will seek out a live person when they want one, based on their own needs and ideas. Most of us would rather not be forced into a conversation by inadequate access to key product information. Investment in learning from the one-on-one conversations we have, and adding to the public knowledge base founded on that learning, pay off in freeing up time to have more interesting conversations. I've seen reductions of up to 75 percent in support e-mail traffic simply by creating informative lists of frequently asked questions (FAQs) and making people aware of them at the point where they're likely to be scratching their heads over a particular type of question. Making certain, of course, the new content isn't written in corporatespeak!

I try to spread the burden of dealing with customer conversations throughout an organization. I make everyone spend some time answering questions from customers. Not only do I mine everyone's budget for the

There are two kinds of FAQs, I realized recently: (1) those that frame questions the company wants you to ask (Q: So how good ARE your products anyway? A: Very very VERY good!!!), and (2) those that acknowledge actual problems and provide solutions. The (1) variety is bullshit PR, while (2) is truly useful.

—Christopher Locke

support costs (let's face it, by the time I'm trying to find money to answer the mail, it's too late) but I also give everyone involved a tap into our customers' heads. It results in a lot more shared awareness of our mission, strengths, and opportunities.

Silence Is Fatal

ONLINE MARKETS WILL TALK ABOUT COMPANIES WHETHER companies like it or not. People will say whatever they like, without caring whether they're overheard or quoted—in fact, having one's views passed along is usually the whole point. Companies can't stop customers from speaking up, and can't stop employees from talking to customers. Their only choice is to start encouraging employees to talk to customers—and *empowering* them to act on what they hear. Freed from restrictions perceived as an unwelcome straitjacket, and are ultimately unenforceable anyway, workers can generate enormous goodwill as everyday evangelists for products and services they've crafted themselves, and thus take genuine pride in.

I've spent the last two years "bringing fire to the cavemen" in the corporate world (specifically fashion). They still don't understand the difference between a server, a browser, and content, but they do understand that they have to be online. They think that a computer is just like a television. That you can just scan some glossy print ads and throw them up on a site. No one cares about usability. No one cares about being real.

They're scared that if they don't make the jump to Internet marketing/selling that they'll lose their customer loyalty. They're scared that if they do make the jump to Internet marketing/selling that they'll lose their customer loyalty. They ask themselves: "Is this really us? Is technology part of our lifestyle/branding concept?" They say they don't market to that demographic, but they know deep down inside that if they don't soon they won't have anyone to market to.

Sometimes I just want to scream.

—Kimberly Peterson, *e-mail to cluetrain.com*

"Customer loyalty" is not a commodity a company owns. Where it exists at all—and the cases in which it does are rare—loyalty to a company is based on respect. And that respect is based on how the company has conducted itself in conversations with the market. Not conversing, participating, is not an option. If we don't engage people inside and outside our organization in conversation, someone else will. Start talking.

CHAPTER

Markets Are Conversations

DOC SEARLS AND
DAVID WEINBERGER

*When you think of the Internet,
don't think of Mack trucks full of
widgets destined for distributorships,
whizzing by countless billboards.*

Think of a table for two.

@man

IT WAS APRIL IN PARIS, SEVERAL WEEKS BEFORE A BIG PRESS
conference where my client, a large but rapidly shrinking French computer company, would roll out a wonderful new computer, the first of its kind. The whole
project had been veiled in secrecy for years. Security was intense. Code names
were used. Deep alliances with Big Players were mentioned only in hushed tones.
The company had hired me to develop a strategy for the rollout. In particular,
they wanted a "message," one that would serve as a tagline for the event and
for all the advertising to follow. A meeting of the company's marketing communications people was convened for my analysis of the market and a briefing on a
strategy that would make the press conference great.

The assignment was painfully hopeless. Oh, the new computer was nice and
the usual customers would buy it, but the larger market—the one this company

needed to penetrate—could care less. The company had been too silent too long. With nothing to lose, I told them the truth.

"We have three problems," I began. "First, there is no market for your message, least of all among journalists, who want facts and stories. Second, there is no market for your secrecy. You have long ignored the market, now they will choose to ignore you. Third, there really is no market for your press conference. Journalists want to be briefed exclusively."

They stared at me. I continued:

"Markets are nothing more than conversations. See these magazines? They're a form of market conversation. We should already be in their stories. We are key to the subject, but we're missing in action after working in secret for years. Our only hope is to talk. Starting now."

I outlined a strategy for igniting as much conversation as possible in a very short time, suggesting some fun, creative, and ultimately pointless ideas. Later, a dozen people came up and thanked me for telling the truth and giving them new hope (although presumably for their next jobs).

Then the project manager took me aside and said, "That was brilliant. Now, what's the tagline?"

First Things Last

THE FIRST MARKETS WERE MARKETS. NOT BULLS, BEARS, or invisible hands. Not battlefields, targets, or arenas. Not demographics, eyeballs, or seats. Most of all, not consumers.

The first markets were filled with people, not abstractions or statistical aggregates; they were the places where supply met demand with a firm handshake. Buyers and sellers looked each other in the eye, met, and connected. The first markets were places for exchange, where people came to buy what others had to sell—and to talk.

The first markets were filled with talk. Some of it was about goods and products. Some of it was news, opinion, and gossip. Little of it mattered to everyone; all of it engaged someone. There were often conversations about the work of hands: "Feel this knife. See how it fits your palm." "The cotton in this shirt, where did it come from?" "Taste this apple. We won't have them next week. If you like it you should take some today." Some of these conversations ended in a sale, but don't let that fool you. The sale was merely the exclamation mark at the end of the sentence.

Market leaders were men and women whose hands were worn by the work they did. Their work was their life, and their brands were the names they were known by: Miller, Weaver, Hunter, Skinner, Farmer, Brewer, Fisher, Shoemaker, Smith.

For thousands of years, we knew exactly what markets were: conversations between people who sought out others who shared the same interests. Buyers had as much to say as sellers. They spoke directly to each other without the filter of media, the artifice of positioning statements, the arrogance of advertising, or the shading of public relations.

These were the kinds of conversations people have been having since they started to talk. Social. Based on intersecting interests. Open to many resolutions. Essentially unpredictable. Spoken from the center of the self. "Markets were conversations" doesn't mean "markets were noisy." It means markets were places where people met to see and talk about each other's work.

Conversation is a profound act of humanity. So once were markets.

The Industrial Interruption

THE ADVENT OF THE INDUSTRIAL AGE DID MORE THAN just enable industry to produce products much more efficiently. Management's approach to production and its workers was quickly echoed in its approach to the market and its customers. The economies of scale they were gaining in the factory demanded economies of scale in the market. By the time it was over we had forgotten the one true meaning of the market, and replaced it with industrial substitutes.

In THE THIRD WAVE, Alvin Toffler wrote that the rise of industry drove an "invisible wedge" between production and consumption, a fact Friedrich Engels had noticed over one hundred years earlier. As production was ramped up to unheard-of rates, the clay pot of craftwork was broken into shards of repetitive tasks that maximized efficiency by minimizing difference: interchangeable workers creating interchangeable products.

In the market, consumption also needed to be ramped up—not just to absorb the increased production of goods, but also to promote people's willingness to buy the one-size-fits-all products that rolled off mass-production lines. And management wasted little time noticing the parallels in efficiencies they could achieve all along the production-consumption chain. If products and workers were interchangeable, then interchangeable consumers began to look pretty good too.

The goal was simple. Customers had to be convinced to desire the same thing, the same Model-T in any color, so long as it's black. And if workers could be better organized through the repetitive nature of their tasks, so customers were more easily defined by the collective nature of their tastes. Just as management developed a new organizational model to enhance economies of scale in production, it developed the techniques of mass marketing to do the same for consumption.

So the customers who once looked you in the eye while hefting your wares in the market were transformed into consumers. In the words of industry analyst Jerry Michalski, a consumer was no more than "a gullet whose only purpose in life is to gulp products and crap cash." Power swung so decisively to the supply side that "market" became a verb: something you do *to* customers.

In the twentieth century, the rise of mass communications media enhanced industry's ability to address even larger markets with no loss of shoe leather, and mass marketing truly came into its own. With larger markets came larger rewards, and larger rewards had to be protected. More bureaucracy, more hierarchy, and more command and control meant the customer who looked you in the eye was promptly escorted out of the building by security.

The product of mass marketing was the *message*, delivered in as many forms as there were media and in as many guises as there were marketers to invent them. Delivered locally, shipped globally, repeated inescapably, the business of marketing devoted itself to delivering the message. Unfortunately, the customer never wanted to take delivery.

The Shipping View

DURING THE INDUSTRIAL AGE, THE MOVEMENT OF materials from production to consumption—from flax to linen and from ore to musket—was a long and complicated process. Potentially vast markets had potentially vast distribution needs. The development of new transportation systems eased the burden, and global systems flourished. Even huge distances could be spanned so that products could be delivered efficiently. Inexorably, business began to understand itself through a peculiar new metaphor: *Business is shipping*. In this shipping metaphor—still the heart and soul of business-as-usual—producers package *content* and move it through a *channel, addressed* for delivery down a *distribution system*.

The metaphor was effectively applied not just to the movement of physical goods, but also quickly applied to the packaging and delivery of marketing content. It's no surprise that business came to think of marketing as simply the delivery of a different type of content to consumers. It was efficient to manage, one size could fit many, and the distribution channel—the new world of broadcast media—was more than ready to deliver. The symmetry was perfect. The production side of business ships interchangeable products and the marketing side ships interchangeable messages, both to the same market, the bigger and more homogeneous, the better.

One problem: *There is no demand for messages*. The customer doesn't want to hear from business, thank you very much. The message that gets broadcast to you, me, and the rest of the earth's population has nothing to do with me in particular. It's worse than noise. It's an interruption. It's the Anti-Conversation.

That's the awful truth about marketing. It broadcasts messages to people who don't want to listen. Every advertisement, press release, publicity stunt, and giveaway engineered by a Marketing department is colored by the fact that it's going to a public that doesn't ask to hear it.

Marketers felt this truth in their bones, and learned to cloak their messages, to disguise them as entertainment, to repackage the content as regularly as business learned to vary this year's product line. Today, we all know and have come to expect this. We are even disappointed if it's not well done. Commercials disguise themselves as one-act plays, press releases play the part of important stories, and advertising masquerades as education. Marketing became an elaborate game between business and the consumer, but the outcome remained fixed. As sophisticated as marketing became, it has never overcome the ability of people to smell the BS behind all the marketing perfume.

It is not hard to understand, then, that "business is shipping" at times felt more like "business is war," another pervasive metaphor. We *launch* marketing campaigns based on *strategies* that *target* markets; we *bombard* people with messages in order to *penetrate* markets (and the sexual overtones here shouldn't be dismissed either). Business-as-usual is in a constant state of war with the market, with the Marketing department manning the front lines.

Consider the distance we've come. Markets once were places where producers and customers met face-to-face and engaged in conversations based on shared interests. Now business-as-usual is engaged in a grinding war of attrition with its markets.

No wonder marketing fails.

The Axe in Our Heads

EVERY ONE OF US KNOWS THAT MARKETERS ARE OUT TO get us, and we all struggle to escape their snares. We channel-surf through commercials; we open our mail over the recycling bin, struggling to discern the junk mail without having to open the envelope; we resent the adhesion of commercial messages to everything from sports uniforms to escalator risers.

We know that the real purpose of marketing is to insinuate the message into our consciousness, to put an axe in our heads without our noticing. Like it or not, they will teach us to sing the jingle and recite the slogan. If the axe finds its mark we toe the line, buy the message, buy the product, and don't talk back. For the axe of marketing is also meant to silence us, to make conversation in the market as unnecessary as the ox cart.

Ironically, many of us spend our days wielding axes ourselves. In our private lives we defend ourselves from the marketing messages out to get us, our defenses made stronger for having spent the day at work trying to drive axes into our customers' heads. We do both because the axe is already there, the metaphorical embodiment of that wedge Toffler wrote about—the one that divides our jobs from our lives. On the supply side is the producer; on the demand side is the consumer. In the caste system of industry, it is bad form for the two to exchange more than pleasantries.

Thus the system is quietly maintained, and our silence goes unnoticed beneath the noise of marketing-as-usual. No exchange between seller and buyer, no banter, no conversation. And hold the handshakes.

When you have the combined weight of two hundred years of history and a trillion-dollar tide of marketing pressing down on the axe in your head, you can bet it's wedged in there pretty good. What's remarkable is that now there's a force potent enough to actually start loosening it.

Here's the voice of a spokesperson from the world of TV itself, Howard Beale, the anchorman in Paddy Chayefsky's NETWORK who announced that he would commit suicide because "I just ran out of bullshit." Of course, he had to go insane before he could at last utter this truth and pull the axe from his own head.

As Seen on TV

I t isn't just working consumers who suffer from axe-in-head disease. CEOs have the same problem. Even hip e-CEOs like Bob Davis of Lycos, Inc. "We're a media company," Davis tells PC WEEK. "We make our money by delivering an audience that people want to pay for." Note the two different species here: *audience* and *people*. And look at their qualities. One is "delivered." The other pays. One is cargo and the other is money. Never mind that we're talking about the *Web* here. Customers—and advertisers—are human. Consumers are not.

Networked Markets

THE LONG SILENCE—THE INDUSTRIAL INTERRUPTION OF the human conversation—is coming to an end. On the Internet, markets are getting more connected and more powerfully vocal every day. These markets want to talk, just as they did for the thousands of years that passed before *market* became a verb with *us* as its object.

The Internet is a *place*. We buy books and tickets *on* the Web. Not *over*, *through*, or *beside* it. To call it a "platform" belies its hospitality. What happens on the Net is more than commerce, more than content, more than push and pull and clicks and traffic and *e*-anything. The Net is a *real place* where people can go to learn, to talk to each other, and to do business together. It is a bazaar where customers look for wares, vendors spread goods for display, and people gather around topics that interest them. It is a conversation. At last and again.

In this new place, every product you can name, from fashion to office supplies, can be discussed, argued over, researched, and bought as part of a vast conversation among the people interested in it. "I'm in the market for a new computer," someone says, and she's off to the Dell site. But she probably won't buy that cool new laptop right away. She'll ask around first—on Web pages, on newsgroups, via e-mail: "What do you think? Is this a good one? Has anybody checked it out? What's the real battery life? How's their customer support? Recommendations? Horror stories?"

"I'm in the market for a good desk dictionary," says someone else, and he's off to Amazon.com where he'll find a large number of opinions already expressed:

> I love the look of this book, and the publisher did a great job; but I made the mistake of buying it without realizing that it was first published over 7 years ago....

> I've had this book for two days and I keep going back to it. I may not be typical since I collect dictionaries and wanted this when I heard about it last year, but....

> Ugh, they don't have "aegritudo" but they have the "modern" definition of "peruse."...

These conversations are most often about value: the value of products and of the businesses that sell them. Not just prices, but the market currencies of reputation, location, position, and every other quality that is subject to rising or falling opinion.

It's nothing new, in one sense. The only advertising that was ever truly effective was word of mouth, which is nothing more than conversation. Now word of mouth has gone global. The one-to-many scope that technology brought to mass production and then mass marketing, which producers have enjoyed for two hundred years, is now available to customers. And they're eager to make up for lost time.

More ominous for marketing-as-usual is this: Finding themselves connected to one another in the market doesn't enable customers just to learn the truth behind product claims. The very sound of the Web conversation throws into stark relief the monotonous, lifeless, self-centered drone emanating from Marketing departments around the world. Word of Web offers people the pure sound of the human voice, not the elevated, empty speech of the corporate hierarchy. Further, these voices are telling one another the truth based on their real experiences, unlike the corporate messages that aim at presenting what we can generously call a best-case scenario. Not only can the market discover the truth in the time it takes to do a search at a discussion archive, but the tinny, self-absorbed voices of business-as-usual sound especially empty in contrast to the rich conversations emanating from the Web.

What's more, networked markets get smart fast. Metcalfe's Law,* a famous axiom of the computer industry, states that the value of a network increases as the square of the number of users connected to it—connections multiply value exponentially. This is also true for conversations on networked markets. In fact, as the network gets larger it also gets *smarter*. THE CLUETRAIN COROLLARY: the level of knowledge on a network increases as the square of the number of users times the volume of conversation. So, in market conversations, it is far

---

* Bob Metcalfe is the inventor of Ethernet, the founder of 3Com, and an INFOWORLD columnist, among many other vocations.

easier to learn the truth about the products being pumped, about the promises being made, and about the people making those promises. Networked markets are not only smart markets, but they're also equipped to get much smarter, much faster, than business-as-usual.

Business-as-usual doesn't realize this because it continues to conceptualize markets as distant abstractions—battlefields, targets, demographics—and the Net as simply another conduit down which companies can broadcast messages. But the Net isn't a conduit, a pipeline, or another television channel. The Net invites your customers in to talk, to laugh with each other, and to learn from each other. Connected, they reclaim their voice in the market, but this time with more reach and wider influence than ever.

When Push Comes to Suck

THE RELUCTANCE OF BUSINESS-AS-USUAL TO BREAK OUT of its set way of thinking was perhaps epitomized best by the Web's own infatuation with "push technology." This reached its zenith in May 1997, when WIRED, the computer industry's utopian fashion monthly, boldly declared its wish to supplant the Web with media more suited to advertising. In its customary overstatement and retinal-torture colors, the magazine devoted its cover and following eleven pages to "PUSH! Kiss your browser Goodbye: The radical future of media beyond the Web." According to the article, the Web was already too demanding for the average spud, so WIRED wanted your inner couch potato to enjoy "a more full-bodied experience that combines many of the traits of networks with those of broadcast. SEINFELD viewers know what we're talking about," the authors wrote.

Ever since the Web showed up, business-as-usual has desperately tried to pipe-weld it onto the back end of TV's history. The money at stake is huge. McCann-Erickson reports more than $45.5 billion spent on TV advertising in the United States alone in 1998. In the same year, total worldwide advertising expenses passed $400 billion. That'll keep a lot of axes in a lot of heads.

But it won't work on the Web, because networked markets aren't passive spectators waiting to receive the next marketing message. The Web isn't home to advertising-as-usual. The "push" movement of 1997 became the pushover of 1998.

The Market That Talk Built

THE POWER OF CONVERSATION GOES WELL BEYOND ITS ability to affect consumers, business, and products. Market conversations can make—and unmake and remake—entire industries. We're seeing it happen now. In fact, the Internet itself is an example of an industry built by pure conversation.

The process of building the Internet was a little like building a bridge: Start with a thin wire spanning a chasm, then spin that single wire into a thick cable capable of supporting heavy girders and the rest of the structure. Incredibly, no one directed this effort. No one controlled it. The people who incrementally built the Internet—literally, one bit at a time—participated solely out of enthusiasm, an enthusiasm driven by a shared and growing vision of what this strange thing they were building might ultimately become.

What if the task of building the Internet had been jobbed out to the leaders of the communications business: to online services like AOL and Compuserve, to network companies like Novell and 3Com, to telecom companies like AT&T and Northern Telecom, to software companies like Microsoft and Lotus?

It never would have happened. It certainly never would have been imagined as it now exists. Every one of those companies would have looked for a way to control it, to make it theirs. More than a few would have turned down the job. Microsoft was famously late to the Internet game in part because Bill Gates thought there was no money to be made.

What it took was behind-the-scenes work by what amounts to a loosely organized, Internet-mediated software craft guild. The results include Apache, a Web server developed by Brian Behlendorf and a bunch of other hackers, simply because they needed it. Today more than half of all the Web's pages are served by Apache.

In fact, nearly a third of the world's Web servers are powered by Linux, the dark-horse challenger to Microsoft's previously unquestioned software hegemony. Linux was initiated by a young, unknown software developer, Linus Torvalds. He needed it, so he crafted it—and then he made it available to the rest of the world through the Internet. He published not just the finished product but, far more important, its source code. Anyone with software engineering tools and the technical chops could add to it, modify it, craft it into precisely the tool they needed. As a result, Linux has rapidly become one of the most sophisticated, powerful, and configurable software products in history—all without anyone managing or controlling it.

Eric Raymond, in his seminal work on hacker culture, THE CATHEDRAL AND THE BAZAAR, describes the dynamics of this distributed and self-motivated community of independent programmers. How was it possible that a seemingly disorganized, seemingly undirected band of renegade hackers could rise to such prominence and threaten the world's largest, most powerful high-tech corporation with the only credible alternative not only to Windows NT, but even to Windows itself?

By conversation. Both the Internet and Linux are powerful demonstrations of a pure market conversation at work. They show what can happen when people are able to communicate without either the constraints of command-and-control management, or the straightjacket of one-message-fits-all. As Raymond writes:

> The thing about the Internet is you can't coerce people over a
> T-1 line, so power relationships don't work.... So the only game
> left to play is pure craftsmanship and reputation among peers.
> If you can offer people the chance to do good work and be seen
> doing good work by their peers, that's a really powerful motivator.

The most important lesson Linux hackers teach is that *whole markets* can rapidly arise out of conversations that are independent not only of business, but also of government, education, and other powerful but hidebound institutions, thanks in large measure to something hackers helped invent precisely for that purpose: the Internet.

Conversation may be a distraction in factories that produce replaceable products for replaceable consumers, but it's intimately tied to the world of craft, where the work of hands expresses the voice of the maker. Conversation is how the work of craft groups proceeds. And conversation is the sound of the market where creators and customers are close enough to feel each other's heat.

What's more, these new conversations needn't just happen at random. They can be created on purpose. "We hackers were actively *aiming* to create new kinds of conversations outside of traditional institutions," Raymond says. "This wasn't an accidental byproduct of doing neat techie stuff; it was an explicit goal for many of us as far back as the 1970s. We *intended* this revolution."

Nice job.

New Messages for Marketing

SO, IF MARKETS ARE CONVERSATIONS (THEY ARE) AND there's no market for messages (there isn't), what's marketing-as-usual to do? *Own* the conversations? Keep the conversations *on message*? Turn up the volume until it drowns out the market? *Compete* with the new conversations?

But how could it? People are talking in the new market because they want to, because they're interested, because it's fun. Conversations are the "products" the new markets are "marketing" to one another constantly online. Hey! Come look at my Web site. Subscribe to my e-zine. Check the whacked-out rant I just posted to alt.transylvanian.polarbears. Get a load of this stupid banner ad I just found at boy-are-we-clueless.com!

By comparison, corporate messaging is pathetic. It's not funny. It's not interesting. It doesn't know who we are, or care. It only wants us to buy. If we wanted more of that, we'd turn on the tube. But we don't and we won't. We're too busy. We're too wrapped up in some fascinating conversation.

Engagement in these open, free-wheeling marketplace exchanges isn't optional. It's a prerequisite to *having* a future. Silence is fatal.

So what becomes of marketing? How do companies enter into the global conversation? How do they find their own voice? Can they? How do they wean themselves from messaging? What happens to

- PR
- advertising
- marketing communications
- pricing
- positioning

...and the rest of the marketing arsenal?

Excellent questions.

Private Relations

IRONICALLY, PUBLIC RELATIONS HAS A HUGE PR PROBLEM: people use it as a synonym for BS. The call of the flack has never been an especially honorable one. There is no Pulitzer Prize for public relations. No Peabody, Heismann, Oscar, Emmy, Eddy, or Flacky. Like all besieged professions, PR has its official bodies, which do indeed grant various awards, degrees, and titles. But do you know what they are? Neither do most PR people. Say that you're an award-winning PR person and most people will want to change their seats.

Everyone—including many PR people—senses that something is deeply phony about the profession. And it's not hard to see what it is. Take the standard computer-industry press release. With few exceptions, it describes an "announcement" that was not made, for a product that was not available, quoting people who never said anything, for distribution to a list of people who mostly consider it trash.

Dishonesty in PR is *pro forma*. A press release is written as a plainly fake news story, with headline, dateline, quotes, and all the dramatic tension of a phone number. The idea, of course, is to make the story easy for editors to "insert" in their publications.

But an editor would rather insert a crab in his butt than a press release in their publication. The disconnect between supply and demand could hardly be more extreme. No self-respecting editor would let a source—least of all a biased one—write a story. And no editor is in the market for a thinly disguised advertisement, which is the actual content of a press release.

Editors hate having to deconstruct press releases to find just the facts, ma'am. To most editors, press releases are just pretend clothing for emperors best seen naked—because naked emperors make much better stories than dressed-up ones.

PR folks are paid to hate stories, even though stories are precisely what the press—PR's "consumers"—most wants. The fundamental appeal of stories is conflict, struggle, and complexity. Stories never begin with "happily ever after," but press releases always do, because that's the kind of story PR's real market—the companies who pay for public relations—demands. The PR version of the *Titanic* story would be headlined 705 DELIGHTED PASSENGERS ARRIVE AFTER THE *TITANIC*'S MAIDEN VOYAGE. Page two might mention some "shakedown glitches inevitable whenever a magnificent new ship is launched." Releases have no room for the very elements that might actually interest a journalist.

Public relations not only fails to comprehend the nature of stories, but

The Real Role of Press Releases

Press releases serve some functions other than pretending to attempt to get ink:

- They let a company try on and then decide on a positioning: "Here's how our story would appear in the press."

- They tell your boss that you're doing a bang-up job: "We shipped the Rangolator 2000 on time, and it's the first uniquely revolutionary enterprise Web-based portal solutionizer. See, it says so right in this fake newspaper article, boss!"

- They give management the illusion that they're going to be quoted in a magazine article.

It'd be nice if companies figured out that they can do all that without actually wasting editors' time with the damn press releases.

imagines that "positive" stories can be "created" with press conferences and other staged events. John C. Dvorak, PR scourge of long standing, says, "So why would you want to sit in a large room full of reporters and publicly ask a question that can then be quoted by every guy in the place? It's not the kind of material a columnist wants—something everybody is reporting. I'm always amazed when PR types are disappointed when I tell them I won't be attending a press conference."

"PR Types." We all know what that means: They're the used car salesmen of the corporate world. You can't listen to PR Types without putting on your highest-grade, activated-carbon bullshit filter. If you're a journalist, you are seen by PR Types as prey. They hunt you down at work, at social events—hell, if you're donating a kidney and a PR Type is on the next table, she'll chat you up about the new product announcement until her anesthetic kicks in and then a little bit longer. Damn PR Types.

But, of course, the best of the people in PR are not PR Types at all. They understand that they aren't censors, they're the company's best conversation-alists. Their job—their *craft*—is to discern stories the market actually wants to hear, to help journalists write stories that tell the truth, to bring people into conversation rather than protect them from it. Indeed, already some companies are building sites that give journalists comprehensive, unfiltered information about the industry, including unedited material from their competitors. In the age of the Web where hype blows up in your face and spin gets taken as an insult, the real work of PR will be more important than ever.

Advertising vs. Word of Web

FAIRFAX CONE, ONE OF THE GREAT MEN OF ADVERTISING, said his craft was nothing more than "what you do when you can't go see somebody." This simple distinction draws a perfect line between TV and the Web. TV is the best medium ever created for advertising. The Web is the best medium ever created for sales. The Web, like the telephone, is a way you can go see somebody, a way to talk with them, show your wares, answer their questions, offer referrals, and make it easy for them to buy whatever they want. Why get someone to look at an ad on the Web when, with exactly the same amount of wrist power, you can get them into your electronic storefront itself?

Sure, you can advertise on the Web, and many Internet companies say advertising is how they are going to make their money. And the sum of advertising on the Web keeps going up. Why not? Just liquidate a few percent of those moon-high stock valuations and buy a few billion dollars more Web advertising. Forrester Research reports that "despite cries that online ads don't work, spending for Internet advertising will continue to grow at a furious pace." They say spending will explode from $2.8 billion in 1999 to $33 billion in 2004.

But Web advertising is already an inside joke. Most of the banner ads you see at the tops of pages are trades and sponsorships, not paid advertising. And everybody knows that having your page turn up in the top ten results when someone goes hunting at a major search site is far more effective than buying ads on Web sites. (This, predictably, has sparked the buying of ads on *search* sites.)

There's no denying that a saturation ad campaign that puts your company's name in tens of millions of banner ads will buy you some name recognition. But that recognition counts for little against the tidal wave of word-of-Web. Look at how this already works in today's Web conversation. You want to buy a new camera. You go to the sites of the three camera makers you're considering. You hastily click through the brochureware the vendors paid thousands to have designed, and you finally find a page that actually gives straightforward factual information. Now you go to a Usenet discussion group, or you find an e-mail list on the topic. You read what real customers have to say. You see what questions are being asked and you're impressed with how well other buyers—strangers from around the world—have answered them. You learn that the model you're interested in doesn't really work as well in low light as the manufacturer's page says. You make a decision. A year later, some stranger in a discussion group asks how reliable the model you bought is. You answer. You tell the truth.

Compare that to the feeble sputtering of an ad. "SuperDooper Glue—Holds Anything!" says your ad. "Unless you flick it sideways—as I found out with the handle of my favorite cup," says a little voice in the market. "BigDisk Hard

Drives—Lifetime Guarantee!" says the ad. "As long as you can prove you oiled it three times a week," says another little voice in the market. What these little voices used to say to a single friend is now accessible to the world. No number of ads will undo the words of the market. How long does it take until the market conversation punctures the exaggerations made in an ad? An hour? A day? The speed of word of mouth is now limited only by how fast people can type. Word of Web will trump word of hype, every time.

Ads may still have hypnotic, subliminal effects, like those tunes we can't get out of our heads (a legacy of the old advertising industry adage "if you have nothing to say, sing it"), but we now have the world's largest support group encouraging us to take that first step: We acknowledge that there is a power greater than ourselves, and it's not some freaking banner ad or a cola company whacking our head with a jingle. It's the conversation that is the Web.

Sites of Salt

YOU MIGHT THINK MARKETING COMMUNICATIONS departments talk about communications. Not really. They actually spend most of their days thinking about how to hide what's really going on in the organization. That's what crafting "messages" is mostly about. For every "message," there are dozens or hundreds of facts—interesting, useful facts—that never get said. Numbers that change. Divisions that move. Features added and subtracted. And that's not counting all the outright negative stuff: the merger that failed, the layoffs, the departed leaders, the stopgap products.

In the Industrial Age—the age of scarce and mostly nonconversational media—there were legitimate reasons for being "on message." The biggest was the need to say one positive thing to everybody at once, in a form that worked equally well in a thirty-second ad and a thirty-page white paper to reach the broadest common denominator.

Even at their most complete—in the form of brochures and other stiff-necked paper goods—marketing communications painted a glossy picture no one believed. We all have been trained by a lifetime of experience to turn down the volume when confronted with a beautiful full-color artifact explaining

Dunk This!

As the Internet shifts the power from merchants to customers you can hear the timbers creaking. For instance, David Felton was mightily—perhaps a tad irrationally—cheesed off when a Dunkin' Donuts shop refused to put skim milk into his coffee, even though the sign in the window said you can get it "your way" there. So he did what any red-blooded Internet denizen would do: He made fun of Dunkin' Donuts on his Web site. And then he grabbed www.dunkin-donuts.org and turned it into a public comment board.

The comments poured in—predominantly negative. Some seemed not worth the calories it took to type them: "I would like to know why the DD on the Fellsway in Medford requires that I buy at least 5 munchkins...if I wanted 5 munchkins I would get a donut." Others were a bit more biting: "I live quite close to [a particular store] and only smell donuts being baked about once a week..." And then there's the employee who warned people to stay away from anything filled with jelly since the employees "toss it around, play with it."

The five thousand franchise owners jumped into the fray. Said Felton in a press release: "It was a common occurrence to see a comment about a particular franchise submitted in the morning, and by the same afternoon, the owner of the franchise already contacted the author and submitted a follow-up on the Web site." But that wasn't enough for the Corporate Big Boys with their big swingin' jelly sticks. They sued the little guy, even though he was getting only fifty to seventy visitors a day. Felton sold the site to Dunkin' for an undisclosed sum. He says he couldn't face the legal bills and harassment. One of the terms is that Dunkin' host a public comment area on its own site. As of this writing, it's not clear that they will post the comments or just suck them into the "Customer Service department," where they will be processed into a crusty new fudge topping prepared by employees who don't wash their hands.

But, if Dunkin' Donuts lacks the munchkins to let the public speak, the public will find another site to do it: www.dunkindonutssucks.com is already registered.

why the products are perfect, the company loves its customers, and every customer is delighted. Like editors skimming a press release, customers root through brochures to find a few motes of useful information. We took all of "marcom's" goods with more than a grain of salt. We needed a whole salt mine to keep up with the tide of BS.

Predictably, most corporate Web sites look like brochures. Visitors have to click through screen after screen of fatuous self-praise to find the few bits of useful information they really want. At least printed brochures don't take as long to download.

If you want to take your first baby step towards entering the market conversation, torch any brochureware on your site. At best your networked market views it as a speed bump, at worst as an insult.

That doesn't mean that you should put up a site that consists of nothing but the facts expressed in Times Roman text (although useful facts are a great place to start). Your site needs to have a voice, to express a point of view, and to give access to helpful people inside your corporation. Replace the brochures with ways to ignite dialogues. Not only do your customers want to talk with real people inside your organization, but your employees are desperate to talk with real customers. They want to tell them the truth.

They will in any event, because your wall of brochures is as solid as a line in the sand.

Fair Market Price

TRADITIONALLY, MARKETING DEPARTMENTS ENGAGE IN pricing exercises to discover a market's ceiling. This makes obvious sense when the supply side controls the means of both production and distribution. But after the revolution, comrade, the old regime's pricing strategies are the first to be led to the gibbet. After decades of replaceable products, replaceable workers, and replaceable consumers, we now have replaceable *merchants*. Think of this as the mass market's revenge.

The first effect of this shift in power has been tremendous downward price pressure. After being trained so assiduously in the economics of mass-ness, the first impulse on the Web is to shop on price alone. Shopping "bots" can find the lowest price among all merchants doing business on the Web. I can go to www.InvoiceDealers.com and see a head-to-head comparison of how little over the invoice price my local car dealers are willing to sell me a new Honda. Or if I decide to buy an Epson Stylus 900 color printer (I have already listened in on the consumer conversations on the Web), I can go to a site like www.computeresp.com and get a list of forty-four merchants, sorted by price, that will sell me one for prices ranging from $330.95 to $404.37. Some of the merchant names may be familiar—Egghead ($346.39) and Gateway ($364.95) —but how much is name recognition worth given that whatever service I may need is going to require the same trip to the post office anyway?

Driving margins towards zero isn't a good thing. Businesses have to make money, after all. And in a war of margin slicing, the Big Boys are often able to stand the heat longer (although A&P managed to burn itself to the ground in the 1970s by initiating a competition with smaller grocery stores to see who could price the furthest under cost for the longest). But it's early yet. And merchants are smart. They offer new services that will distract the market from its insistence on extracting vengeance by shaving margins with a guillotine. And what are those emerging services, hmm? *Conversations.*

For example, the merchant may enable you to talk with its own experts. Or it may put you in touch with the rest of the market directly, using the means the Web has served up to us. Amazon.com famously presents readers' reviews and rankings. For technical support, Microsoft directs you to Usenet-style discussion groups, which it's smart enough not to try to control.

In short, although there is no demand for messages, there is a tremendous demand for good conversation. That's one way merchants fight commoditiza-tion. But both no-margin pricing and higher-margin pricing with the added value of conversation are still examples of pricing driven from on high as if suppliers were still in charge. Increasingly, they're not. In fact, in the most exciting new markets developing on the Web, the demand side—the market— tells the suppliers just what they're willing to pay. This is quite literally true at www.priceline.com, where you can let hotels, airlines, mortgage companies, and car dealerships know exactly what your best offer is. They can take your business or leave it. That's up to them. But the pricing is up to you.

The most dramatic move away from top-down pricing is evident at auction sites such as www.ebay.com, which enable the market to sell to itself. Yes, eBay is a virtual flea market, albeit it with millions of items on sale at any one moment. But it is also much more. Soon after eBay took off, some merchants realized that they don't need a Web storefront of their own; they can just offer their wares at the auction site. They lose some control over pricing—in a complete role reversal, the market sets the price and the merchant has the power to say no—but they get into the thick of the fray with almost no startup or marketing costs. And it's not just consumers who are engaged in auctions. PNC Bank Corp. in Pittsburgh accepts bids on interest rates for certificates of deposit. Deere & Co. auctions used farm equipment. Ford Motor has auctioned automotive components.

Now the Web is reaching even further up the chain, fundamentally changing the value and cost equations that rationalize pricing. With the music encoding standard called MP3, any digital recording—such as tracks from a CD—can be posted on a Web site, downloaded, listened to, and even recorded back onto a CD. Everyone who knows how to point and click can gather tracks from their favorite musicians and assemble their own albums. Production and distribution are so cheap and easy that the market can do it for itself. That leaves the recording industry with almost nothing but the role of marketing, a task they generally haven't grasped very well when it comes to the Web because they're too busy trying to squelch what they rightly see as a threat to their hegemony. Recording companies thought they were originators but instead found they were intermediaries. And the most efficient markets tend to have the fewest intermediaries.

Which brings us to the top of the chain: in this case, the musicians themselves.

Why would musicians allow their music to be downloaded for free rather than sold for fifteen to twenty dollars by a recording company? Maybe because it's a good marketing technique for selling CDs and concert tickets. Maybe because they hope that fans will eventually be willing to pay them something—

much less than the price of a typical CD—for the download privilege, just as shareware has proved a successful business model for many software developers. Maybe musicians will allow their music to be priced low enough to encourage the widest possible distribution because they are *craftspeople* who care more deeply about the value of their work than its price. And maybe it's because they define that value in terms other than what they charge for one form of finished goods.

Waiting for e-Books

I t doesn't take a genius to see that what MP3 is doing to the music business, the widespread availability of reading matter on the Web will do to the publishing industry. It's already happening. And it will take off like wildfire the moment truly readable, portable computer displays enter the mainstream.

And that is marketing's final pricing challenge. Pricing interchangeable products for a mass market is just a matter of testing how high you can raise the bait out of the water and still have the fish bite. Set the price, maybe tweak it, and you're done: All the fish are going to have to pay the same price. But when it comes to prices, the Web acts like a craft world in which prices aren't uniform across all the products. Each hotel room, each Beanie Baby, and each hand-assembled CD can now be priced according to different rules, granting the customer new advantages. The mechanical transactions in which the price declared by the supplier was paid by the consumer now becomes more of a dance, sometimes a courtship, and always a conversation.

Assume the Position

Every morning when I wake up,
I try to remember who I am and where I come from.

Harry S. Truman

PUBLIC RELATIONS, ADVERTISING, AND MARCOM ALL reflect the company's "position." Positioning is darned important, then. Strategic, even. And if you're a marketing consultant, positioning is where the big bucks are. You're right there at the top of the marketing totem pole.

Positioning is not only lucrative for its practitioners, it's also fun, since it's usually done on a blank piece of paper. "Who do we want to be?" asks the positioning expert. "Are we the maker of the world's finest timepieces? No, maybe we're the people who keep business on time. Ooh, maybe we're the company that's making punctuality into a fashion accessory!" Undoubtedly, someone will trump these suggestions by saying, "We're not really about watches at all," and then, in a solemn voice: "We're the Time Company."

Often, "positioning exercises" become expensive sojourns into corporate psychology. The consultant gets to spend time with one group leader after another, performing the role of corporate shrink. The resulting data are impossible to connect, but that doesn't matter, because the goal is only to come up with a "statement." And all that statement has to be is marginally different from every other company's faked-up statement. Never mind that nobody in the marketplace gives a damn about any company's positioning statement. It only matters that *this* statement will "drive the strategy," which will be yet another advertising and PR bombing campaign.

Can it get more arrogant? Well, actually, yes.

Positioning wasn't even an issue until 1972, when Al Ries and Jack Trout wrote a series of articles for ADVERTISING AGE and then authored one of the top-selling business books of all time, POSITIONING: THE BATTLE FOR YOUR MIND. The goal of positioning, Trout says, is to own *one word* in your customer's mind. For evidence, you don't even need to leave your own skull. Take a look: You'll find Fedex in the "overnight" position, Crest in the "cavities" position, and Volvo in the "safety" position, even if you never buy those products. In the battlefield of your mind, those companies are entrenched in those positions.

Why one word? Because to Trout and Ries, the human mind is as closed as a clam and just as roomy. Witness Jack Trout's "five basic principles of the mind," from THE NEW POSITIONING:

1. Minds are limited.

2. Minds hate confusion.

3.   Minds are insecure.

4.   Minds don't change.

5.   Minds lose focus.

In short, minds are so pathetic that they desperately need help, even if it comes in the form of an axe. That's what positioning is for.

Too bad, because positioning actually is about something much more important, something that gets trivialized by those who reduce it to generating a catchy tagline. Positioning is about discovering who you, as a business, are— *discovering* your identity, not inventing a new one willy-nilly. Positioning should help a company become what it is, not something it's not (no matter how cool it would be).

A company can certainly try to be what it's not. But the market conversation will expose the fakery. One clue is any attempt by a company to deny its history, because history is one of those things that just can't be changed. GM will always be the product of Alfred Sloan's preference for implementation over innovation, Apple will always come from Steve Jobs's artistic temperament, Hewlett-Packard will always come from its founders' obsession with quality products for niche technical markets, Nordstrom will always come from the family's original shoe business.

Of course companies and products can change their identities (and even their natures) over time. Volkswagen no longer bears (for most of us) the history stated in its very name: Hitler's car for the proud German people. Kellogg's Razzle Dazzle Rice Krispies no longer bear much connection to the obsessive health concerns of the company's founder. But such changes generally are gradual and often painful. In fact, if they are too rapid and too easy, the market conversation will be merciless in exposing the phoniness it sniffs.

There are other clues that a company is having an identity crisis:

1.   Is there a spark of life in its marketing materials? Do they smack of focus groups and the safety of the lowest common denominator, or do they take the risk of being as interesting as its best customers?

2.   Do its marketing programs keep people out or invite them in?
Do they help customers and prospects make connections to
the relevant employees?

3.   Is the company able to admit a mistake? Can employees admit
they disagree with management decisions or the latest marketing
mantra? Or must they always explain why everything is perfect
in this, the best of all possible companies?

4.   Is the company so jealous of its "image" that it has surgically
implanted a lawyer where its sense of humor used to be?

5.   Does it drill its employees on the corporate catechism, or can the
workers tell stories that for them capture the essence of what the
company is about?

6.   Do the employees routinely sign their e-mail "Views expressed
do not necessarily reflect those of the management"?

These indicators have a common theme. Each points to a gap between who
your company is and what it says it is. The gap is where inauthenticity lives,
and the exposure of the gap constitutes corporate embarrassment. Much of
marketing is devoted to papering over that gap. Deming gave the deathless
advice: "Drive out fear." We might add: and drive out shame.

But how can a business be authentic? Authenticity describes whether someone truly owns up to what she or he actually is. Since corporations and businesses aren't individuals, ultimately their authenticity is rooted in the employees. If the company is posing, then the people who *are* the company will have
to pose as well. If, on the other hand, the company is comfortable living up to
what it is, then an enormous cramp in the corporate body language goes away.
The marketing people won't create throwaway lines that are clever but false.
The sales folk will walk away from the "sales opportunities" that the company
is better off losing than having to support. The product developers won't pro-
pose features that look good on paper but do their customers no real good.

None of this has to do with one-word positioning statements, press release
boilerplate, or pledging allegiance to corporate goals before every company
meeting. It has to be learned in the heart, not by rote. What we learn through

memorization affords us no spontaneity. We can recite the right words, but they're not our own—we can't riff on them. The market conversation can spot marketing recitatives within two syllables because the Web thrives on spontaneity. We are all so tuned to the sound of the real human voice that, given a chance to interact, we can't be fooled...at least not for long.

And if a company is genuinely confused about what it is, there's an easy way to find out: Listen to what your market says you are. If it's not to your liking, think long and hard before assuming that the market is wrong, composed of a lot of people who just are too dumb or blind to understand the Inner You. If you've been claiming to be the Time Company for two years but the market still thinks of you as the Overpriced Executive Trophy Watchmaker, then, sorry, but that's your position. If you don't like what you're hearing, the marketing task is not to change the market's *idea* of who you are but *actually* to change who you are. And that can take a generation: Look at Volkswagen.

Entering the Conversation

THE CHAPTERS ON PR, ADS, MARCOM, PRICING, POSITIONING —hell, all of them—in THE MARKETING-AS-USUAL MANUAL OF STRATEGY AND TACTICS need to be redone. It's not because the war has shifted from the air to the ground, or because now we're fighting guerrillas instead of massed troops. No, marketing-as-usual thinks it's fighting a war when in fact the "enemy" is having a party: "Hey, dude, put on this Hawaiian shirt, grab some chips and dip, and join in. But first you gotta loosen your grip on that assault weapon."

Here's some advice on entering the conversation: Loosen up. Lighten up. And shut up for a while. Listen for a change. Marketing-as-usual used to be able to insert its messages into the mind of the masses with one swing of its mighty axe. Now messages get exploded within minutes. "Spin" gets noticed and scorned. Parodies spread ad campaigns faster than any multimillion-dollar advertising blitz. In short: The Internet routes around a-holes.

So, enter the conversation and do it right. But how?

Corporate Voice

HERE'S A SYLLOGISM. YOUR COMPANY NEEDS TO ENGAGE in the new market conversations. Conversations occur in human voices. Your voice is the public expression of your authentic identity, of who you really are, of where you really come from. So let's draw the logical conclusion: On the Net at least, your company can't engage in the market conversation without its authentic voice.

Sounds simple. But what does it mean when applied to a corporation? Corporations don't have voices. They don't have mouths to speak with, or hands to type with, or body language to betray their real intentions, or eyebrows to punctuate a joke. Corporations are legal fictions.

But *businesses* aren't fictions. Businesses are as real as families and nations. As with all social entities, they speak as the sum of the parts, as the individuals who *are* the parts, and everything in between.

A business has a voice. You can usually hear it—authentic or unauthentic— most obviously and transparently, on its Web site. Even before the last graphic finishes downloading, you can usually tell if the company speaks with passion, if it's lost or uninterested, or if it's online just because some consultant said it has to be. You can tell if the business has some perspective on itself or whether it's all wrapped up in being the Number One Provider of Something, Anything, Please! You can tell if it wants to talk with you or just to pick your pocket. You can tell if the people who work there really care or if they always carry their résumé with them, just in case. You can tell if the company is basically lying or basically telling the truth.

Funny

The Web is a funny place—literally. We learn a lot about the voices we hear through their humor. Having a sense of humor tells us that you have some distance from yourself and the dreadful seriousness of your concerns. It is, in most cases, a prerequisite for personal authenticity—if you're not laughing at yourself, are you really being honest about yourself? After all, you are—like all of us—a ridiculous creature.

Ah, but can you *really* tell? All the customer has to go by are bits on a screen. Couldn't a clever marketing person pony up a page that looks hip and happy, successfully masking the cries of anguish coming from the corporate cube farm?

Yes, for a while. Marketing has been training its practitioners for decades in the art of impersonating sincerity and warmth. But marketing can no longer keep up appearances. People talk. They get on the Web and they let the world know that the happy site with the smiling puppy masks a company with coins where its heart is supposed to be. They tell the world that the company that promises to make you feel like royalty doesn't reply to e-mail messages and makes you pay the shipping charges when you return their crappy merchandise. The market will find out who and what you are. Count on it.

That's why you poison your own well when you lie. You break trust with your own people as well as your customers. You *may* be able to win back the trust you've blown, but only by speaking in a real voice, and by engaging people rather than delivering messages to them.

The good news is that almost all of us already know how to talk like real people. It's just a matter of pulling that fat axe from our skulls.

The Wrong Kind of Buzz

> *English is the perfect language*
> *for preachers because it allows you to talk*
> *until you think of what to say.*
>
> **Garrison Keillor**

IT'S EASIER TO LOCATE AND DISARM THE MARKETING messages buzzing in our heads than to disable the vocabulary that's been slipped in. At the word level, we all at times slip into the old marketing-speak. Nowhere is this more true than in the technology industries. For example, Bob Epstein, back when he ran Sybase, once gave an otherwise good speech in which he used the expression "extended enterprise client server." Afterwards a number of attendees were asked if they could recall this phrase. Most could remember that the phrase was a bunch of buzzwords, but none could remember the phrase itself.

This is because "extended enterprise client server" is composed entirely of TechnoLatin, a vocabulary of vague but precise-sounding words that work like the blank tiles in Scrabble: you can use them anywhere, but they have no value. TechnoLatin takes perfectly meaningful words and empties them. If language is a living organism, TechnoLatin words are like those pod people in the movie INVASION OF THE BODY SNATCHERS. They look real, but they are not. And like the pod people, TechnoLatin has become the norm. Clarity is the exception when it should be the rule. Today we no longer make chips, circuit boards, computers, monitors, or printers. We don't even make products. Instead we make *solutions*, a fatuous noun further bloated by empty modifiers such as *total*, *full*, *seamless*, *industry standard*, and *state-of-the-art*.

Equally vague and common are *platform*, *open*, *environment*, and *support* when used as a verb. A veterinarian using TechnoLatin might say that a dog serves as a platform for sniffing, is an open environment for fleas, and that it supports barking.

This isn't language. It's camouflage.

A perfect example of TechnoLatin's mindless power is a press release that heralded the pointless name change of the semifamiliar Xymos to the anonymous Appian Technology:

> Over the past two years, Xymos has been repositioning itself. No longer a typical semiconductor supplier, the company has focused on its ability to integrate advanced technologies that use innovative system architecture and software into high performance system solutions for PCs and workstations.

If communication had taken place here, we would probably know what Appian Technology now does for a living. But because the release is written in TechnoLatin, it offers no such clues. While Xymos was at least "a typical semiconductor supplier," Appian Technology isn't even a noun. Instead it is "focused" on an "ability" to "integrate" a pod salad of "advanced technologies," "innovative system architecture," "high performance system solutions," and so forth.

Since "Appian" was first a famous Roman highway, you'd think this might be a clue to Xymos's new identity. But the release says:

> Appian was chosen for the name because it represents the ability
> to use leading edge technology and innovation, integrated into
> solutions that provide differentiation and competitive advantage.

Just what the Romans had in mind.

The obligatory quote from Appian's president and CEO really hits the nail on the board: "What we have done at Appian Technology is couple leading-edge technology with innovation, and integrated it into high performance system solutions which provide customers with differentiation and competitive advantage." This took two years?

Amazingly, Appian Technology did not kill itself. Instead it quietly yawned into a coma. Today its stock maintains a newsless flat line at 1/128th above zero.

It's obvious why we fall into TechnoLatin even if we know better. We sound so smart when we use words no one quite understands. We sound so precise. And we sound like we *belong*: "Distributed platform environment" does for technology marketers what "you know, like, whatever..." does for teenagers.

And, of course, it's not just the technology industry that's in love with pod words. Brochureware at www.ford.com talks about the Lincoln's "advanced performance characteristics," "leading-edge safety systems," and "AdvanceTrac™ yaw control." Not to be outdone, Honda says its Odyssey (the car Odysseus would have chosen to drive to Troy, no doubt) has a "rear crush zone," an "advanced Traction Control System," and "Grade Logic programming." And, of course, restaurants have their own cant (call it GastroFrench, followed by Nouveau GastroFrench), as do interior decorators, sportscasters, Boy Scout leaders, and just about everyone else—loose-limbed phrases that are trotted out as if the real words of the craft were somehow too humble.

So our advice: Speak real words. The new Web conversations are remarkably sensitive to the empty pomposity that has served marketing so well. Until now.

Who Speaks?

BUT WHO GETS TO SPEAK?

Companies feel a tremendous urge to control communications; it seems as bred-in-the-bone as wanting to sell products. They create org charts to define who gets to do the talking. They issue policy statements: Only PR can talk to the press. Only Investor Relations can talk to financial folk. Only the CEO can talk to THE JOURNAL. We can't afford to muddy our message or dislocate our positioning. God knows what some disgruntled worker might tell valuable customers! So, let's set up a command hierarchy and station it in a hardened communications bunker.

You might as well try to sew closed a fishing net. The simple fact is that your employees are already joining the market conversation. And in most cases it's because they find conversations about what they are working on to be really interesting. They like talking with customers. They like to help. And, sorry to point this out, but they also like complaining if the business is flawed at heart.

The one thing they *don't* want to do, would never do on their own, is deliver a message. And if you make that their role, they will be exposed immediately as company tools. We're all superb at sniffing out the shills: They lack spontaneity, their language is stilted, and they are just a little *too* happy. In fact, in Usenet newsgroups, it's not uncommon to find participants being warned about postings from particular people who've blown their credibility by sounding like corporate mouthpieces.

So, what's a business to do? People aren't going to simply repeat messages. You can't shut them up—at least not for long—and you can't make them mouth words they don't believe any more than you could get your teenaged children, your spouse, your friends, or *anyone* to. Save your discipline for the few renegades who, through malice or ignorance, spill beans that need to be kept in the can. Expend your efforts instead on building a company that stands for something worthwhile, so that you can't wait to unleash every single one of your voices into the wilds of the new global conversation.

The Web of Voices

BUT WHAT ABOUT THE RISK? SUPPOSE A "LOWLY CLERK" speaks for the company in public and says something wrong? Something action-able? Something confidential, or sensitive? Lordy, what would become of us then?

Let's put this differently: Shall we agree to let the sun rise tomorrow? It's going to happen. It already is happening. And it's always happened. The mail clerk describes the corporate strategy to the stranger next to him on the bus, and then provides a critique. The technical-documentation writer tells her cousin how to circumvent the cover-your-ass "safety" lid. And the telephone support rep tells a customer—on company time!—that one of the features touted on the box doesn't really work exactly as described.

Each of these people is speaking for the company. But, through a game of selective attention, businesses claim their unauthorized personnel aren't really speaking for the company. Not *officially*. Officially there are communication channels that generally correspond to the corporate hierarchy. Officially the entire corporation speaks through a single orifice. Anything that issues from it is sanctioned, true, and legally actionable. Anything that does not come through "approved channels" is just the random lip-flapping of employees for which the corporation is not legally liable.

Of course, all of the excitement, all of the heat, all of the jazz comes from these flapping lips who are speaking for the corporation in everything but the legal sense. They're improvising, not staying "on message." They're pursuing the interests they share with the customers, not corporate interests that are at war with customers. Businesses that try to get their people to say exactly the same thing in exactly the same words ("No, Jenkins, for the hundredth time, you've left 'enhanced' out of 'The world's leading manufacturer of enhanced software solutions for maximizing supply chain advantage'!") are losing their greatest marketing resource. Now those lips have the global megaphone of the Web. In a Webbed world, loose lips *float* ships.

The people to whom employees are talking and have always talked are sophisticated enough to know that there's a big difference between a Saturn technician answering a customer question in a discussion on the Web and an official reply from Saturn technical support. They know the contents of a press release are not the same as the pep talk that Saturn's president gives at the company picnic. Part of listening to a voice is assessing the role of the words and the speaker. The Web is giving us lots of training in that.

There is, of course, the legal question. While the people engaged in a conversation almost always know precisely the degree of official-ness with which someone is talking, lawyers worry that someone could unintentionally or maliciously take a casual remark as official policy. And of course that could happen. But businesses take legal risks just by shipping products.

Framejacking

I use framejacking to refer to getting contexts radically wrong: throwing garbage into a trash can that's for sale in a hardware store, or asking George Clooney to look at your child's rash. Marketing-as-usual has to fight its desire to engage in some serious framejacking on the Web. For example, at a conference, I once heard a representative of a national bank explain with pride that his bank sends out e-mails touting services targeted to particular users, under the signature of the local manager. The messages artfully include typos. And they make sure not to send them out at 2 A.M. since that would strain the credulity of their pigeons, um, customers. This is a distressing spin on "1:1 marketing," a phrase invented by Don Peppers and Martha Rodgers to describe providing personalized offers and services from a mass marketer. Since it's really software that's addressing the "1" customer, this should probably be called 0:1 marketing and, when it pretends to be really 1:1, it's framejacking.

Likewise, personalization software enables sites to modify what they show you based upon your history of interactions there. This can make a site much more useful to each visitor, but if it tries to make it seem that behind the site is a real human being who loves you for yourself, well, that's some serious framejacking...and a serious betrayal of trust. —David Weinberger

For that reason, it'd be good for employees to make every effort to clarify the status of their remarks. No, this does *not* mean that they should sign their e-mail with the phrase "Does not necessarily represent the views of Management, etc.,"—a common sign of worker alienation. The real aim is to communicate the status, not to introduce still more legalisms. Or a business could forbid its employees from talking on the Web during business hours and from identifying themselves as employees after hours. They could build a firewall that—to use the more apt metaphor—turns their company into a black hole on the Web. Of course, the Web conversation would go on without a hiccup. When the company's silence becomes noticeable in some discussion—"Hey, why doesn't someone from ABC Corp. explain how to keep its product from catching on fire if you put the key in upside down?"—the void will be filled by expert (but exasperated) customers, and then by competitors. If the company truly succeeds in turning itself into a black hole, it may indeed not be talked about on the Web. Or anywhere.

There's your risk for you.

Robo-Voice

Technology is clearing away some of the non-conversational conversations we used to have: "Hello. How are you? Fine, thanks. What can I do for you today? Is that it then? OK, you have a good day" repeated twelve times an hour, forty hours a week. An ATM will do just fine, thanks. Yes, we lose some human contact, and there are certainly places where the brief dialogue with the bank teller is part of a social fabric that shouldn't be sundered— part of the ancient marketplace that has survived. If that's true for you, then you'll avoid ATMs. But in many instances, automated transactions are liberating people from semi-automatic human behaviors.

In other cases, the automation afforded by the Web is *increasing* the quality of offline marketplaces. For example, in part to compete with Amazon.com, local bookstores are enhancing their conversational accoutrements, and many of us appreciate them all the more precisely because of the "frictionless" ease of acquiring books from Amazon.

On the positive side, by acknowledging that, inevitably, many people speak for a particular company in many different ways, the company can address one of the most important and difficult questions: How can a large company have conversations with hundreds of millions of real people?

First, the conversations don't all have to be truly interactive. Few people insist on personal service immediately from every Web site. We're delighted to look through the Frequently Asked Questions (FAQs) and ReadMe files to find our answers—answers culled from the real interactions between the company and its market.

When a conversation is required, or even just desired, being able to count upon a rich range of corporate spokespeople is crucial. That's the only way a growing business can satisfy the market's demand for conversation. For example, at Western Digital's Web site (http://www.wdc.com), users can post technical questions about the company's hard disk drives. Most of those questions are, naturally, about drives that don't work. A Western Digital support person will post an answer, often within hours, and the entire exchange is open to public view, unfiltered. As a result, customers with problems can usually find a previous exchange that answers their questions. Sure, visitors to the site find out that not all Western Digital drives work flawlessly forever, but this is hardly a news flash. More important, they learn that Western Digital has enough confidence in its products to let customers air their gripes, and that if a drive breaks, it'll get diagnosed and fixed at the speed of the Web. And, perhaps most important, they see that the company's customers and enthusiasts care enough to dive in. Could a company ask for better living testimonials? Could customers ask for a livelier, more reality-based source of information?

If you want to hear the sound of the new marketing, listen to these conversations coming from inside, outside, over, and above even the hardest-shelled companies that still think marketing means lobbing messages into crowds. Here is the same sound our ancestors heard in those ancient marketplaces, where people spoke for themselves about what mattered to them.

How to Talk

WE'RE ALL LEARNING TO TALK ANEW. WE'RE ALL GOING to get it right *and* get it wrong. Two events in the fall of 1994 still serve as good cases in point for crisis management. In one case, resolute Ivory Tower isolation caused a major disaster. In the other, real conversation among concerned individuals saved the day.

First, an anatomy of a disaster. Through the 1980s and early 1990s, Compuserve hosted many of the best online forums. One of these professional salons was the EETIMES Forum, hosted by ELECTRONIC ENGINEERING TIMES, the top magazine for the people who design and work with computer chips. It's a safe bet that most of the participants used Intel-based computers, and engineered computers with "Intel Inside." Yet when news of a bug in an early Pentium chip was first found and posted on the forum, nobody seemed to take it too seriously. They joked about it a bit, but took it in stride. After all, bugs in chips are nothing new. But all of them clearly were looking for Intel to jump in and talk about it.

However, there was radio silence from Intel until Alex Wolfe, an EETIMES reporter, wrote about the bug in his magazine. Soon the major media picked up the story and all hell broke loose.

To deal with this crisis, Intel CEO Andy Grove posted something on the forum that read like a papal encyclical on how Intel works. This included a description of a caste system that drew a line between those who should be concerned about such a bug and those who should not, and offered to replace the defective chips for the first group. This didn't sit well with anybody, but the forum members were tolerant at first. They wanted to get to the bottom of this thing, so they attempted to engage Grove on the matter. After all, he had shown up. He must have been willing to talk. But it quickly became obvious that Grove was just posting a notice—the big guy was not going to take part in a conversation.

So, when Intel got shellacked in the press, little help came from what should have been company friends in the engineering community. After all, these were Intel's real customers. They understood how bugs happen. They were articulate and authoritative. But they were just as silent for Grove as

Grove had been for them. Intel was publicly embarrassed into recalling every one of the defective chips, and estimates for reputation damage ran into many millions.

Meanwhile, over in Compuserve's Travel Forum, another bad PR event was taking shape. This one involved United Airlines, which was experiencing a bumpy take-off with its new Shuttle By United service. Like the EETIMES Forum, the Travel Forum had serious participants: high-mileage fliers, pilots, air traffic controllers, travel agents, and airline personnel from every level.

If you could hook up a meter to the forum and measure goodwill, the needle reading for Shuttle By United at take-off was way over on the negative side. Luggage was being lost (three times for one passenger). Passenger loading was chaotic. Customers were unhappy.

Then one United worker (one of those "owners" United's ads talked about so much at the time) jumped in and simply started to help out. The response was remarkable. Here are a few examples:

> **"Good to see someone at United interested!"**
>
> **"Nice to have a UAL person to chat with...thanks."**
>
> **"As a 100k flier, I'm glad to see one of you online here."**
>
> **"I am a pilot for United and I thank you for taking the time to answer all of these questions about the Shuttle."**
>
> **"Nice to see a UA employee on and participating instead of just lurking."**
>
> **"As a UA 1K FF [top-grade frequent flyer] and a PassPlus holder I appreciate your time and interest in the forum."**
>
> **"Don't leave United. You're important to us. Your comments are helpful. You make a difference."**

This kind of conversation moved the meter all the way over to the positive side, just because one company guy took on the burden of talking with customers and trying to solve their problems. *One guy.*

Then one day the same UA employee posted a notice that said, "Due to a conflict with corporate communication policies at United Airlines of employees responding to issues of any nature without the explicit direction of the Communications Division, I will not be participating any longer. I hope this situation changes in the future. Until then, direct any concerns to the Consumer Affairs department at United's World Headquarters."

You can imagine what followed. United got flamed royally by their employee's new friends on the forum.

But, unlike Intel, United stayed in the conversation. A United higher-up jumped in and quickly communicated United's willingness to learn this new form of market relations. The original United correspondent and the higher-up both stayed in the conversation and started to work things out. The needle went back over to the positive side. And nobody ever heard bad news about the Shuttle By United bug.

Lessons learned? The party's already started. You can join or not. If you don't, your silence will be taken as arrogance, stupidity, meanness, or all three. If you're going to join, don't do it as a legal entity or wearing your cloak of officialdom. Join it as a person with a name, a point of view, a sense of humor, and passion.

Marketing Craft

THE MARKET STARTED OUT AS A PLACE WHERE PEOPLE talked about what they cared about, in voices as individual as the craft goods on the table between them. As the distance between producer and consumer lengthened, so grew the gap between our business voice and our authentic voice. Marketing became a profession, an applied science, the engineering of desirable responses through the application of calibrated stimuli—including the occasional axe in the head.

Marketing isn't going to go away. Nor should it. But it needs to evolve, rapidly and thoroughly, for markets have become networked and now know more than business, learn faster than business, are more honest than business, and are a hell of a lot more fun than business. The voices are back, and voice brings craft: work by unique individuals motivated by passion.

What's happening to the market is precisely what should—and will—happen to marketing. Marketing needs to become a craft. Recall that craft-workers listen to the material they're forming, shaping the pot to the feel of the clay, designing the house to fit with and even reveal the landscape. The stuff of marketing is the market itself. Marketing can't become a craft until it can hear the new—the old—sound of its markets.

By listening, marketing will re-learn how to talk.

The Hyperlinked Organization

DAVID WEINBERGER

BUSINESS SOUNDS DIFFERENT THESE DAYS.

The words at meetings have an edge. The language used with customers is unfiltered. The e-mails are pithy, often piercing.

Where once bombastic self-confidence got you taken seriously, now being funny does.

In fact, there's laughter everywhere, although it signals insight and bitterness as often as delight.

Sometimes you could swear you hear children's laughter in the background, over the sounds of cooking and cockatiels and the UPS truck arriving.

Beneath the formalities of business—the committees, the schedules, the payroll checks, the spray of assignments falling from above—there's a buzz, no, the sound of twigs breaking underfoot as paths are trod on the way to human connection. The most amazing thing: You can tell who's talking by listening to the voice.

People are beginning to sound like themselves again.

Intranet Apocalypso

YOU MAY NOT HEAR ANY OF THIS AT YOUR PLACE OF WORK.
But if the Web has touched your business—and it has—then the sound is there.

The odd thing is that you almost certainly have to be making some of the
new sound to hear it. Otherwise, it passes for noise, like an overtone of the
60-cycle thrum of modern business at its automated, time-slicing best.

You hear it or you don't. You get it or you don't. The gulf that has opened
in companies is about the size of the human heart.

That's what makes the situation so ripe for humor. And anger. And absurdity.

CONSIDER THIS: FROM THE OTHER SIDE OF THE GULF
opened by the Web, virtually all of the structures that management identifies
as being the business itself seem to be bizarre artifacts of earlier times, like
wearing a powdered wig and codpiece to the company picnic.

The gulf the Web opens is, ironically, that of connection. Without anyone
asking for it, the Web has given the people inside an organization easy access
to one another in a rich variety of ways. They can send e-mail to one person,
to a steady group, to a dynamic team, to the entire sales force, or "just" to the
board of directors. They can post creative, informative pages that express their
interests, correct the mistakes in the official technical documentation, or point
to the industry analyst's report the company doesn't want anyone to read. They
can write a 'zine that parodies the company line savagely and without let-up.
They can play backgammon online or blow up their colleagues in a ruthless
game of Quake in which the guy who never speaks at meetings routinely turns
his manager into animated meat chunks. They can also find every piece of
information about the company and its competitors, shop for a car, or learn
how to play the blues like Buddy Guy.

The Web, in short, has led every wired person in your organization to
expect direct connections not only to information but also to the truth spoken
in human voices. And they expect to be able to find what they need and do

what they need without any further help from people who dress better than they do. This has happened not because of a management theory or a best-selling business book but because the Web reaches everyone with a computer and a telephone line on her desk.

So, the gulf opens between those who are connected and those who think an office with a door is a sign of success. The gulf is one of expectations, and expectations always guide perception. As a result, the company thinks it's doing one thing while accomplishing the direct opposite with its connected employees. For example:

- The company **communicates** with me through a newsletter and company meetings meant to lift up my morale. In fact, I know from my e-mail pen pals that it's telling me happy-talk lies, and I find that quite depressing.

- The company **org chart** shows me who does what so I know how to get things done. In fact, the org chart is an expression of a power structure. It is red tape. It is a map of whom to avoid.

- The company **manages my work** to make sure that all tasks are coordinated and the company is operating efficiently. In fact, the inflexible goals imposed from on high keep me from following what my craft expertise tells me I really ought to be doing.

- The company provides me with a **career path** so I'll see a productive future in the business. In fact, I've figured out that because the org chart narrows at the top, most career paths necessarily have to be dead ends.

- The company provides me with all the **information** I need to make good decisions. In fact, this information is selected to support a decision (or worldview) in which I have no investment. Statistics and industry surveys are lobbed like anti-aircraft fire to disguise the fact that while we have lots of data, we have no understanding.

- The company is **goal-oriented** so that the path from here to there is broken into small, well-marked steps that can be tracked and managed. In fact, if I keep my head down and accomplish my goals, I won't add the type of value I'm capable of. I need to

browse. I even need to play. Without play, only Shit Happens.
With play, Serendipity Happens.

- The company gives me **deadlines** so that we ship product on
  time, maintaining our integrity. In fact, working to arbitrary deadlines makes me ship poor-quality content. My management doesn't
  have to use a club to get me to do my job. Where's the trust, baby?

- The company looks at **customers** as adversaries who must be won
  over. In fact, the ones I've been exchanging e-mail with are very
  cool and enthusiastic about exactly the same thing that got me
  into this company. You know, I'd rather talk with them than with
  my manager.

- The company works in an **office building** in order to bring
  together all of the things I need to get my job done and to avoid
  distracting me. In fact, more and more of what I need is outside
  the corporate walls. And when I *really* want to get something
  done, I go home.

- The company rewards me for being a **professional** who acts and
  behaves in a, well, professional manner, following certain unwritten rules about the coefficient of permitted variation in dress,
  politics, shoe style, expression of religion, and the relating of
  humorous stories. In fact, I learn who to trust—whom I can work
  with creatively and productively—only by getting past the professional act.

Something's gone wrong. Or maybe something now is starting to go right.

What's wrong isn't trivial. It isn't fixed with dress-down Fridays, health food
in the cafeteria, or learning to pretend to look into the eyes of the trembling
subordinate you're condescending to chat up on the way in from the parking
lot. The power structure, the politics, the sociology, even the spirituality of
work has a sick, sour smell to it.

But you don't need big words. It all begins with pictures. That's why our
hairy-backed ancestors were sketching bison on the wall: They were learning
to see. So let's think instead about the basic *picture* we have of business.

Inside Fort Business

SOMEWHERE ALONG THE LINE, WE CONFUSED GOING TO work with building a fort.

Strip away the financial jibber-jabber and the management corpo-speak, and here's our fundamental image of business:

- It's in an imposing office building that towers over the landscape.

- Inside is everything we need.

- And that's good because the outside is dangerous. We are under siege by our competitors, and even by our partners and customers. Thank God for the thick, high walls!

- The king rules. If we have a wise king, we prosper.

- The king has a court. The dukes, viscounts, and other sub-luminaries each receive their authority from the king. (The king even countenances an official fool. Within limits.)

- We each have our role, our place. If we each do the job assigned to us by the king's minions, our fort will beat all those other stinking forts.

- And then we will have succeeded—or, thinking it's the same thing, we will say we have "won." We get to dance a stupid jig while chanting "Number one! Number one!"

This fort is, at its heart, a place apart. We report there every morning and spend the next eight, ten, or twelve hours inaccessible to the "real" world. The portcullis drops not only to keep out our enemies, but to separate us from distractions such as our families. As the drawbridge goes up behind us, we become businesspeople, different enough from our normal selves that when we first bring our children to the office, they've been known to hide under our desk, crying.

Within this world, the Web looks like a medium that exists to allow Fort Business to publish online marketing materials and make credit card sales easier than ever. Officially, this point of view is known as "denial."

The Web isn't primarily a medium for information, marketing, or sales. It's a world in which people meet, talk, build, fight, love, and play. In fact, the Web world is bigger than the business world and is swallowing the business world whole. The vague rumblings you're hearing are the sounds of digestion.

The change is so profound that it's not merely a negation of the current situation. You can't just put a big "not" in front of Fort Business and say, "Ah, the walls are coming down." No, the true opposite of a fort isn't an unwalled city.

It's a conversation.

Hyperlinks Subvert Hierarchy

FORT BUSINESS'S ASSUMPTIONS ARE BEING CHALLENGED by a meek little thing: a hyperlink.

How could something so small alter the fundamentals of business life? Easy. This wee beastie represents an important change in how pieces are put together —and since all of life is about putting pieces together, this isn't a wee thing at all.

Sure, businesses are legal entities. But that's just a piece of paper. In fact, the real business is the set of connections among people.

Modern business almost universally has chosen a particular type of togetherness: a hierarchy. There are two distinguishing marks of a hierarchy: It has a top and a bottom, and the top is narrower than the bottom. Power flows from the top and there are fewer and fewer people as you move up the food chain.

This not only makes the line of authority crystal clear, it also enhances the allure of success by making it into an exclusive club. As La Rochefoucauld once said, "It is not enough that I succeed. It is also necessary that my friends fail."

No wonder so many of us stare at our bare feet in the morning and wonder why we're putting on our socks.

A couple of other points about business hierarchies:

First, they assume—along with Ayn Rand and poorly socialized adolescents—

that the fundamental unit of life is the individual. This is despite the evidence of our senses that individuals only emerge from groups—groups like families and communities. (You know, it really does take a village to raise a child. Just like it takes a corporation to raise an ass kisser.)

But the Web obviously isn't predicated on individuals. It's a web. It's about the connections. And on the World Wide Web, the connections are hyperlinks. It's not just documents that get hyperlinked in the new world of the Web. People do. Organizations do. The Web, in the form of a corporate intranet, puts everyone in touch with every piece of information and with everyone else inside the organization and beyond.

The potential connections are vast. Hyperlinks are the connections made by real individuals based on what they care about and what they know, the paths that emerge because that's where the feet are walking, as opposed to the highways bulldozed into existence according to a centralized plan.

Hyperlinks have no symmetry, no plan. They are messy. More can be added, old ones can disappear, and nothing else has to change. Compare this to your latest reorganization where you sat down with the org chart and your straightedge and worried about holes and imbalances and neatness for heaven's sake! A messy org chart is the devil's playground, after all.

Second, business hierarchies are power structures only because fundamentally they're based on fear.

Org charts are pyramids. The ancient pharaohs built their pyramids out of the fear of human

Understanding Through Hyperlinks

Ever since Aristotle, we understand what something is by seeing what category it's in and how it differs from other things in that category. For example, humans are in the category of animals, but we are the rational ones. This knowledge hierarchy constrains things to a single category. In a hyperlinked world, however, things can be understood by reference (via metaphor) to other things—and can be like more than one thing at a time.

If you change the way we understand things, you've changed something very significant indeed.

mortality. Today's business pharaohs build their pyramidal organizations out of fear of human fallibility; they're afraid of being exposed as frightened little boys, fallible and uncertain.

To be human is to be imperfect. We die. We make mistakes.

Sometimes we run from our fallibility by being decisive. But doubt is the natural human state, and decisiveness—more addictive than anything you might shoot into your veins—is often based on a superstitious belief in the magic of action.

Within the pyramid we have defined roles and responsibilities. We tell ourselves that this is so the business will run efficiently, but in fact having a role brings us the great comfort of having a turf where we're pretty confident we're not going to be shown up...except maybe by that ambitious jerk on the fourth floor, but we've figured out a way to hook his brains out through his nose, which should delay him at least for a little while.

Of course, dividing the business up into fanatically defended turfs doesn't really protect anyone from fallibility and uncertainty, the very things that mark us as humans.

So, here's some news for today's business pharaohs: Your pyramid is being replaced by hyperlinks. It was built on sand anyway.

THE WEB LIBERATES BUSINESS FROM THE FEAR OF BEING exposed as human, even against its will. It throws everyone into immediate connection with everyone else without the safety net of defined roles and authorities, but it also sets the expectation that you'll make human-size mistakes rather frequently. Now that you've lost the trappings of authority, and you find yourself standing next to the junior graphic designer for gawd's sake, and you can't hide behind your business card, what the hell are you going to do?

You're going to talk with her. You're going to have a conversation. And if you harrumph and try to make sure she knows that you're Very Important by the power vested in you by the power that vested in you, well, she's going to laugh once out loud and five times in e-mail and tell everyone else what an asshole you are.

You see, the hyperlinks that replace the org chart as the primary structure of the organization are in fact conversations. They are the paths talk takes. And a business is, more than anything else, the set of conversations going on.

Business is a conversation because the defining work of a business is conversation—literally. And "knowledge workers" are simply those people whose job consists of having interesting conversations.

"Can I super-size that?" "Have it on my desk by the morning," "There's no *I* in Team," and laughing at your manager's unfunny jokes are not conversations. Conversations are where ideas happen and partnerships are formed. Sometimes they create commitments (in Fernando Flores' sense), but more often they're pulling people through fields of common interest with no known destination. The structure of conversations is always hyperlinked and is never hierarchical:

> To have a conversation, you have to be comfortable being human —acknowledging you don't have all the answers, being eager to learn from someone else and to build new ideas together.

> You can only have a conversation if you're not afraid to be wrong. Otherwise, you're not conversing, you're just declaiming, speechifying, or reading what's on the PowerPoints. To converse, you have to be willing to be wrong in front of another person.

> Conversations occur only between equals. The time your boss's boss asked you at a meeting about your project's deadline was not a conversation. The time you sat with your boss's boss for an hour in the Polynesian-themed bar while on a business trip and you really talked, got past the corporate bullshit, told each other the truth about the dangers ahead, and ended up talking about your kids—that maybe was a conversation.

Conversations subvert hierarchy. Hyperlinks subvert hierarchy. Being a human being among others subverts hierarchy.

B o t t o m - U p

THE WEB IS UNDOUBTEDLY A PART OF YOUR BUSINESS
plans. You've got it safely contained, under control, managed. Why, your organization has probably already installed a corporate intranet so it can publish the
human resource policies that no one read on paper to people who now won't
read 'em on screen. Excellent!

Yes, your centralized corporate intranet has eliminated some paper and is
making management feel vaguely cool. But that's not the web that's going to
shake the foundations of your fort.

While you've been hiring consultants to create a slick corporate intranet,
establishing policies about who gets to post what, and creating a chain of command to ensure that only appropriate and approved materials show up on your
internal corporate home page, your engineers, scientists, researchers—hell even
the marketing folks—have been creating little Web sites for their own use.

No one is controlling what's posted on them except the people doing the
posting. No one is making sure that the corporate logo is in the right place. No
one is making sure that the writing is official, officious, and as dull as the pen-
cil drawer of a recently downsized middle manager.

The real party got under way while you were still setting up the banners at
the corporate prom. (This year's prom theme: "Responsibility in a Web Age!")

For example, by the time Sun Microsystems got around to counting, they
had eight hundred intranets. And when Texas Instruments put in their corporate intranet, they invited everyone who had one already in place to register
with the top-down one. Within a few months, two hundred and fifty internal
sites had registered, and no one knows how many unregistered ones there
were. Even a top-down intranet can take on a bottom-up feel, as happened at
Lucent Technologies, according to an article in THE WALL STREET JOURNAL.
After Lucent brought together a product-development team of five hundred
engineers across three continents and thirteen time zones, it watched dozens
of them insert their own pages into the project intranet. Some of these pages
related directly to the project; others were strictly personal, like, "Hey, look at

this picture of me and my dog!" Either way, the project took on a human cast that never would have been present otherwise. In the end the team leader attributed the success of the project in no small part to "the ultimate Democracy of the Web."

Granted, these are technology companies, but you don't have to be a technical genius to create an intranet. If someone wants to share some information, they can turn their computer into a Web server. It's free, and it's getting easier every day.

The intranet revolution is bottom-up. There's no going back. If a company doesn't recognize this, the top-down intranet it puts in can breed the type of cynicism that results in ugly bathroom graffiti and mysterious golfing cart accidents.

The intranets under the radar screen—and the rest of the Net panoply, including e-mail, mailing lists, and discussion groups—ignore the corporate blather and ass-covering pronouncements. Instead, these new Web conversations are actually being used to get some work done.

The Character of the Web

IT'S WEIRD, BUT NOT TOTALLY UNEXPECTED.

It turns out that the Web is infecting organizations with the characteristics of its own architecture. So, if you want to know what a hyperlinked organization looks like, look at what the Web itself is like.

What's the Web's character? You can slice it into seven basic themes:

1. **Hyperlinked.** Before the Web, computer networks were laid out in advance like well-planned cities. Who got connected to whom and how was all part of the master plan. And once you were connected, there was a recognizable central authority responsible for the whole shebang. The Web isn't even a little like that. The Web literally consists of hundreds of millions of pages hyperlinked together by the author of each individual page. Anyone can plug in and any page can be linked to any other, without asking permission. The Web is constantly spinning itself—many small pieces loosely joining themselves as they see fit.

2. **Decentralized.** No one is in charge of the Net. There is no central clearing house that dispatches all requests and approves all submissions. No one ordered the Web built. There is no CEO of the Web. There is no one to sue. There's no one to complain to. There's no one to fix it when it breaks. There's no one to thank.

3. **Hyper time.** Internet time is, famously, seven times the velocity of "normal" time. And yet we use the leisurely verb *browse* to describe our behavior on the Web because in the virtual world, I feel I can move about at my own pace, exploring when and where I want. I can take a quick look at a site and come back later without having to find another parking space, go to the end of the line, or pay a second entry fee. The Web puts the control of my time into my hands.

4. **Open, direct access.** The Net provides what feels like direct access to everyone else on the Net and to every piece of information that's ever been posted. If you want to go to a page, you just click on the link and, boom, you're there. (The fact that this might have required, beneath the surface, thirty "hops" among servers in places you never heard of is completely irrelevant. You don't see the hops; you just see the page.) There's nothing standing between you and the rest of the world of people and pages.

5. **Rich data.** The currency of the Web isn't green-bar printouts of facts and stats. It's pages. Humans have been creating pages since the invention of paper and dirty water. Pages—or "documents" as we sometimes say—are extraordinarily complex ways of presenting information. Typically, they tell you as much about the author as about their topic, a big change from the pre-Web information environment that aimed at generating faceless data.

6. **Broken.** Because the Web is by far the largest, most complex network ever built, and because no one owns it or controls it, it is always going to be, in the words of Tim Berners-Lee, the inventor of the Web, "a little bit broken."

7. **Borderless.** Because traditional networks were concerned as much with security as with access, it was usually made clear where your stuff ended and other people's stuff began. The Web, on the other hand, was designed so that you can include a link to

a page without having to get the author's permission. Thus, on
the Web it is often hard to tell exactly where the boundaries are.

From these characteristics of the technical architecture of the Web come
the changes that are transforming your business.

The Hyperlinking of the Organization

YOUR ORGANIZATION IS BECOMING HYPERLINKED.
Whether you like it or not. It's bottom-up; it's unstoppable.

Despite the wet stink of fear, you ought to be delighted. Hyperlinked
organizations are closer to their markets, act faster, and acquire the valuable
survival skill of learning to swerve.

Of course, they also are impossible to manage—although they can be
"unmanaged"—and you'll have to give up your pretense of power, status, and
lordliness. But, then, as the old saying has it, you can't make an omelet without
nuking the existing social order.

Here's the drill for the rest of this chapter. We've just discussed seven key
characteristics of the Web. Now we're going to go through them one at a time,
in order (no talking in the hallways and please stand to the right to enable
those in a hurry to pass) to see what's happening inside organizations touched
by the Web—that is, all organizations to one degree or another.

Let's Put the Hyper Back into Hyperlinks

Here's one example of how things work in a hyperlinked organization:

> You're a sales rep in the Southwest who has a customer with a
> product problem. You know that the Southwest tech-support per-
> son happens not to know anything about this problem. In fact,
> she's a flat-out bozo. So, to do what's right for your customer you
> go outside the prescribed channels and pull together the support
> person from the Northeast, a product manager you respect, and a
> senior engineer who's been responsive in the past (no good deed
> goes unpunished!). Via e-mail or by building a mini-Web site on an
> intranet, you initiate a discussion, research numbers, check out
> competitive solutions, and quickly solve the customer's problem—

all without ever notifying the "appropriate authorities" of what
you're doing because all they'll do is try to force you back into
the official channels.

It's a little thing. But it's a big change in the ground rules of work. The official structure is of little use to you. Instead, your network of trusted colleagues becomes paramount. Your effectiveness depends upon how networked you are, how hyperlinked you are.

The hyperlinked teams you form may not be as project-centered as in the example above. As organizations become hyperlinked, they spawn hyperlinked committees, hyperlinked task forces, hyperlinked affiliations, hyperlinked interest groups, hyperlinked communities, hyperlinked cheering squads, hyperlinked pen pals, and hyperlinked attitudes. Humans seem to fill up every available social niche just as nature itself abhors an ecological vacuum.

These hyperlinked relationships are, like the Web of hyperlinked documents, a shifting context of links of varying importance and quality. They are self-asserting, not requiring anyone else's authority to be put in place. And the value of the individual "node" to a large degree depends upon the node's links.

This last point is a big shift. Links have value by pointing *away* from themselves to some other site. All Web pages derive *some* value from the links on them. (A page with no links is literally a dead end on the Web.) In fact, the single most-visited site on the Web, Yahoo!, derives almost all of its value not from what it contains but from what it points to. Yet our understanding of the nature of knowledge, education, and expertise is bound up with things that *contain* value, not with things that point you out of themselves to find value elsewhere. Books get their value from their content. Education is the transfer of content into the receptacle that is the student. And an expert is someone who contains a lot of information, like a book contains information. In fact, experts are people who can write books. But, with today's huge increase in the amount of information, you can be an expert only in something sliced so thin that often it's trivial. Increasingly, a *useful* expert is not someone with (containing) all the answers but someone who knows where to find answers. The new experts have value not by centralizing information and control but by being great "pointers" to other people and to useful, current information.

In short, your most valuable employee is likely to be the one who, in response to a question, doesn't give a concrete answer in a booming voice but who says, "You should talk to Larry. And check Janis's project plan. Oh, and there's a mailing list on this topic I ran into a couple of weeks ago...."

How could you hope to capture this on an org chart? And how do you compensate people fairly if their value depends upon their participation in a shifting set of hyperlinked associations? How do you hire great hyperlinked people? How could this ever be expressed on a résumé?

Great questions...because there aren't clear answers yet. Epochal changes are not Q&A sessions. We're at the beginning of the biggest Q since the Industrial Revolution. It's a time to make things up, try them out, fail a thousand times, and laugh at how stupid you look.

The urge to "solve the problem" is nothing but the voice of the old command-and-control psychosis trying to reassert itself.

("Premature elucidation": the plight of men who come to answers way too soon.)

Decentralizing the Fort

Traditionally, business is an indoor sport.

Businesses by their very nature are centralized (or so we think). Even if you are a global enterprise, your organization consists of a headquarters with regional offices. A business is, after all, a bringing together of talented people who agree to work to achieve some common goals. We've assumed that "together" means we have to centralize power, control, and resources. But there are lots of ways to be together.

Metaphorical Togetherness

We often assume that complex projects can only be accomplished through centralized planning and control. It worked for building the Hoover Dam, after all. Not to mention World War II.

But, of course, it only works for some types of wars in some types of places.

And the builders of the Hoover Dam aimed at creating a massive physical object with delicate dependencies so that there was only one way to succeed and many ways to fail.

A marriage is a far more complex project than any business partnership, and centralized control doesn't work real well there. Raising a family is likewise a complex project that cannot be centrally organized or planned. And, of course, the most complex network ever imagined—the World Wide Web—has been implemented without any central control whatsoever.

But is a business more like a family than a war? Absolutely. Wars are won and then are over, but companies don't declare victory and disband the troops. And although both wars and companies have missions, companies don't ever issue a press release that says:

> OmniCo is proud to announce that on June 23 we accomplished
> our mission of being the world's leading supplier of low-temperature
> oil to the miniature locomotive industry, so we are now demobiliz-
> ing so that our brave men and women can rejoin their families.
> Good night and God bless.

No, families and businesses are open-ended commitments.

Suppose running a business is more like farming than like waging war. Perhaps the real aim of business is to build a place that provides a high yield over the long term, responding to the sometimes vast changes in the environment.

Command and control don't work when you're cultivating the wilderness, when you're experiencing an ecology of surplus, when changes happen faster than response times, when you're homesteading, not marching to battle. Why is it even necessary to have to point out something so obvious?

That's not a rhetorical question. We all know enough about the inequities of history to smell something suspicious about the insistence on centralized control. Control and management are the mantras of the people who are in power, who judge personal success by power, and who use power to keep themselves at the top.

Org charts are written by the victors. But hyperlinks are created by people finding other people they trust, enjoy, and, yes, in some ways love.

Self-Reliance

In a decentralized environment, people figure out that they have to do things themselves. Indeed, they *want* to do things themselves.

This is a well-known phenomenon in customer support: People would rather find the answers themselves on your Web site than have the answers delivered to them by picking up the phone. This may sound like a control issue, but in fact it is about time. By browsing your support Web site, not only can I choose when I'm going to look and how long before I'll give up, I can click through some screens, work on something else, eat lunch, maybe even bookmark the page, and come back to it tomorrow. I can complete the task on my own schedule.

Mission Statements

Mission statements are directed to customers and prospects—that's why they don't contain the words "and make us filthy rich"—but they're also part of a conceptual hierarchy that mirrors the power hierarchy (excuse me, org chart). At the top is the mission statement. And it begat the strategy. And the strategy begat the tactics. And the tactics begat the objectives that begat the tasks that begat the people in cubicles who no longer beget children because they're working all weekend trying to finish the !@#$-ing assignments they've been given to serve the all-powerful mission statement. If you want to know where to find the real corporate point of view and values, look to the stories that are told off hours when folks are "just talking." In short, at the strategic off-site where you're drafting your new mission statement, if you're given a choice between bringing in a consultant or beer, choose the beer.

It shouldn't be surprising that self-reliance is high up in the list of Top Ten Web Virtues. The Web itself started out as a huge do-it-yourself project, and being able to do your own technical support is a mark of Web competency still.

More and more, employees and customers want to feel their own hands on the wheel.

Another obvious example: In the old days, if you wanted to find some information, you had to go to the corporate Information Retrieval Expert and fill out a form. But now the Web has reset expectations. If the data hierophants tell you that you're not trained enough to search the corporate library, you'll reply, "Hey, I just came back from AltaVista,"—or Excite or Hotbot or any of the myriad of search sites—"and I searched through hundreds of millions of pages without any training."

It's time to hand over the keys to the index. Baby's learned to drive.

Self-reliance, however, goes far beyond the technical realm. For example, Boeing enables mechanics to order parts themselves (through their cleverly named Part Analysis and Requirement Tracking—PARTS—system), instead of petitioning the Purchasing department. And Chrysler encourages employees to make their own travel arrangements via an intranet site that shows them only the appropriate choices (for example, the Concorde doesn't show up as a possibility), saving administrative costs and giving the workers a greater sense (illusion?) of control.

There's a dark side to self-reliance. It can encourage a type of arrogant cynicism that reacts to anything that the business tries to do for you with: "I can do it better than that." In this view of the world, there's what I can do with my own two hands and then there's red tape. To the Web cult of self-reliance, the business is not only an obstacle, it's them, the other.

Yet if we know that routing a customer comment through the standard structures of the Fort will result in a content-free form letter being sent out six weeks later, we will sit down and bang out an e-mail immediately that actually addresses the customer's concern. Self-reliance breeds disengagement with the business but more direct engagement with the real work of business.

We are seeing, then, a realignment of loyalties, from resting comfortably in the assumed paternalism of Fort Business to an aggressive devotion to making life better for customers. The business isn't a machine anymore, it's a resource I alone and we together can use to make a customer happy.

Hyper Time

We all know that Internet time is seven times the speed of normal time. ("On the Internet, everyone knows you don't have time to spellcheck.") It affects our business expectations for Internet startups and our expectations about the quality of products we know have been rushed to market, but there's actually more at stake. In fact, the philosopher Martin Heidegger had it right when he 'splained that time is at the root of all that is.

Business likes to think that it operates on a master schedule that devolves into lots of supporting schedules, just as the corporate strategy devolves into objectives and then into tasks, and just as the org chart foliates into branches, twigs, and finally leaves. In a perfectly run business, all the schedules tick in sync. Tick tock tick tock.

Now, we all know that no complex organization works perfectly, so with a knowing smile we dismiss the possibility—but yet we hold it out as an ideal. Our clocks are supposed to be driven from on high.

The Web decentralizes time by letting hyperlinked groups form that are driven by their do-it-ourselves

To: self@evident.com
From: clocke@panix.com

You write: "The business isn't a machine any more, it's a resource, stuff that I alone and we together can use to make a customer happy."

I'm always a little bothered by this making-customers-happy trope. Is this *really* why people try harder in orgs? I tend to doubt it. Maybe it's just me, but I don't think people are too often inspired by customers—that's usually just more happy talk.

Instead workers go the extra mile because *they* want to do a good job. To me, pride of craftsmanship is a far greater motivator. If I'm interacting with a customer, I want to show that other *person* that *I* am a professional even if the rest of the company is going to hell in a handbasket.

And I think this is true right up to the level of the CEO.

If you, David Weinberger, want to make your customers happy, I'd say it's more to demonstrate that you're a capable dude—not because you're concerned for your client's mental state.

To: clocke@panix.com
From: self@evident.com

I think we're probably both right (yech). Having customers who are happy to your face is a very big turn-on for a lot of people. You can say that that's only because ultimately it makes *me* feel competent, but you could reduce *all* human actions to that and thus obliterate the real distinction between stealing candy from a baby and devoting your life to working with the poor.

zeal to get stuff done now. For these groups, schedules are driven locally, not centrally. The schedules are created by local groups and individuals, accounting for their assessment of what's realistic. And they route around obstacles, not like building straight-line highways where it's assumed that all boulders can be blasted out of the way.

Deadlines

But what happens to deadlines if time becomes decentralized?

Let me give you an example that I recount with little pride. I was working in a relatively small software company that was, happily, experiencing growing pains as we went from $3 million to $40 million in revenues. I had been one of the three members of the executive management team that had agreed to roll the dice that set us off on the steep growth curve. My role was strategist. I was never much of an implementer. But because we now desperately needed to run marketing programs, I agreed to step into the role of VP of Marketing. A couple of months later, we hired a Chief Operating Officer to manage our growth. On purpose he was a counter-cultural figure in the company: a hard-bitten, ultra-realistic guy with a relentlessly positive attitude applied as a fresco to mask a cracking wall of disagreeable fear.

A couple of weeks after arriving, he called me into his office to bond with me and also, not incidentally, to find out when the next wave of marketing materials would be ready. I said I didn't know. Why not? he demanded. I replied that I had a really well-motivated team of professionals who were moving heaven and earth to get it all done; it would be done at the earliest possible moment.

He looked at me in amazement. And gave up on me.

Now, I will admit that as COO, he needed to have some sense of the timing of events. For example, he might have needed to know when the materials would be ready because of an upcoming sales meeting. And in such a case I would have told him what I thought would be ready. And if he wanted it sooner, I would have warned him that some of it would be of poor quality. But, in fact, there was no upcoming event. He managed by holding people to deadlines. I managed by holding people to people.

His view of me, to this day, is that I am an unrealistic, soft-edged, namby-pamby, probably borderline homosexual type of guy. My view of him is that he's an unrealistic, anal-retentive, power-driven, frightened little boy. (You know, but underneath it all, we actually don't like each other.)

Is one of us more realistic than the other? I don't think so. If not living by deadlines is unrealistic, it's just as unrealistic to think that a motivated group of people, working hard, will get things done by a particular moment just because you set that moment as the endpoint.

Clearly there's room for both personality types (gosh, I am namby-pamby, aren't I?), but since the deadline drivers always get to state their point of view, let's for once not assume that deadlines are the only way to manage, and that people who miss deadlines are like dawdling children who need to be sent to the corner of the org chart where they can sit to think about what they've done.

Instead, let's leave open the possibility that deadlines are frequently a weapon used by managers who assume that workers are basically slackers. In fact, hyperlinked teams—ruled by the laws of connection—are motivated by a genuine desire to turn out a product or help a customer.

Realism = Pessimism

It's time to take reality back from the bottom-line crowd. When they say "be realistic" they mean "assume the worst, be pessimistic." If someone has underestimated sales potential, no one calls out the ultimate putdown: "Be realistic."

The Web itself was not a realistic proposition: the world's largest network created without management or control? The Web is as implausible as a bumblebee flying. Both are flat impossibilities.

Businesses that define "realism" as "pessimism" will miss all the interesting opportunities.

They will work as hard as they can to do right by their customers and their coworkers. They know better than anyone, in many instances, when the work can realistically be finished. Managing them simply means asking them.

Personal Work Time

The decentralization of time creates other ripples. When you allow people to control their own schedules, they don't always cut their day into clean work and nonwork time periods. Their personal lives begin to invade Fort Business. They know that even if they leave for an hour for the Good News Assembly at their child's elementary school, they still can get done what needs doing, even if it means working at home over the weekend.

Once the time wall is breached, it rapidly becomes more and more permeable. If the only time I can make calls to further the process of adopting a child is during work hours, I will make those calls. If the only time I can talk with a travel agent to plan my vacation is between nine and five, I will call the travel agent during work. And, of course, the Web makes it easier than ever for me to permeate my work time with personal errands and concerns.

Now, the fact is that office workers have always ignored the temporal walls and called the adoption agency and the travel agency during work hours. We've just had to lie and pretend: Sorry to have a life, sir. It won't happen again, sir.

Some businesses have started to recognize that the temporal walls are full of windows. For example, Aetna realized that their workers inevitably spend time at work on personal issues. And what sort of fish-hearted bastard would tell them not to? So, Aetna built into its intranet the sort of information they thought employees were looking for. You can get information about how to adopt a child, for example, or how to arrange for a college scholarship for your kids. Since people are going to spend "business time" doing that anyway, why not make it easier for them by including the information on the corporate intranet?

Yes, doing this had a practical purpose that speaks to the bottom-line guys because it meant employees were spending less time on nonbusiness issues. But you don't need bottom-line reasons to do this type of thing if we take as a basic business principle that companies need to wake up and smell the coffee. The walls around Fort Business may have inscribed in them, "Let all who enter here abandon all personal life," but only the truly pathetic pay it any mind. Therefore, you might as well drop the pretense.

Skimming Time

Let's review, shall we? The Web's decentralization of time breaks apart the master schedule that supposedly has us ticking and tocking in unison, using artificial deadlines to enforce the corporate will. And personal time infects the purity of our time working behind the Fort walls.

One more thing: The Web changes time from sequential to random.

If you already know what "random access" is, feel free to jump over this paragraph (but not before snickering at its witty self-reference). For example, audio is a sequential medium because you can only get from point A to point C by causing point B to pass the tape heads. CDs, DVDs, and hard drives are random access devices because you can hop all around on them.

The Web is (generally) random; you're expected to hyperlink around, sampling what you like. Random access spoils you. Instead of having to wait around for the tape to play out, you can skip right to the parts that you care about. (That also means each person's experience of the tape may be different.)

The Web is making us impatient with anything we can't *skim*. This includes:

- Sixty-seven-slide PowerPoint presentations

- Almost all meetings

- Being put on hold when you place a call that an adequate Web site would have made unnecessary

- Canned online tutorials

- Managers who hand you a copy of a report and then insist on telling you everything that's in it

- Television without a remote

- Traveling

- Bores

The point? Web time isn't just seven times faster than normal time. It's also a thousand times more random—in the good sense.

Open Access to Everything

WHEN IT COMES TO INFORMATION, THE WEB'S IMPULSE is the opposite of Fort Business's. The Fort views information access as a publishing process, pushing to the appropriate people precisely the information they need at the right time. The publisher will ascertain your information needs. Your job is to sit back, relax, and open wide.

This model made sense when information was scarce. And it made sense when business could take itself seriously as an omniscient potentate.

That was then. Now employees want to be able to run barefoot through the tall grass of information. And not simply because we're in a self-reliant sort of mood.

It's one thing to ask an information retrieval specialist to look up some data in a sanctified database where all the data is assumed to be approved and certified. It's another to try to gather competitive information from the Web where you're reading corporate BS from competitors, whining complaints on Usenet, reports from self-proclaimed industry experts (like, um, some of us CLUETRAIN authors), and libelous comments from anonymous stock manipulators.

In this environment, making judgments about what counts is a honed skill, one as personal as writing well or having a sense of humor. It is not something we're willing to delegate to others.

Ah, but the Central Committee says that it must control all access because it can't afford to let out state secrets. Imagine if our competitors got their hands on that stuff!

Sure, there are some trade secrets so important that you need to transport them in briefcases chained to some treasured bodily appendage: Coke's secret formula, a new molecule developed by a bioengineering company, the stocks a mutual fund company is about to invest in. But those are the exceptions. To talk about the role of secrecy in terms of those types of secrets is like evaluating the rural lifestyle by taking Ted Kaczynski's cabin as your example.

And there's a price to assuming that secrecy is normal, that everything is to be kept secret unless otherwise noted. Not only do you have the expense of keeping the secret, but you lose the value of information. Information by its nature only has value insofar as it's known. And, when combined with smart people with an impulse to solve problems and exploit opportunities, information increases its value.

Information wants to be free, sure. But it wants to be free because it wants to find other ideas, copulate, and spawn whole broods of new ideas.

Controlling information is like trying to control a conversation: It can't be done and still be genuine. You're not publishing information, you're building a kitchen, you're planting a field. People wander around in information and learn where to find the stuff that counts, the stuff that's wrong in enlightening ways, the stuff that's purposefully off-base, the stuff that's fun, the stuff that's ludicrous.

The CIA and Secrets

Maybe ten years ago, I heard a tale about the CIA. The agency went on a declassification rampage. It seems that—because the rules of declassification had been so broadly delimited, and the penalties for noncompliance were terrifying—nearly everything for the previous several decades had been classified top secret.

Therefore the agency didn't know what was really top secret anymore. Phone numbers to order-in pizza were filed along with the number of the Kremlin hotline.

Because of fear, the categorization of the information had become practically worthless.

—Christopher Locke

Let's look at one specific type of information that needs to be free: documents.

Heroic Documents

Business currently has a heroic view of documents. When we're given an assignment—"Should we do this merger?" "We need a plan for moving into the new office space"—we go to our cubicle and put our heads down for a day, a week, a fortnight. We go through as many drafts as we have to until we have a killer document—a report or an overhead presentation, typically—that nails it all down, comes to conclusions, and is irrefutable.

Then we go to the big meeting and slap it down like Beowulf slaying Grendl. "Here I stand," we declare, bravado masking our anxiety. And if someone calls our bluff, if someone says, "Hmm, you seem not to have consulted the study the Gartner Group did last quarter" or "You haven't considered the impact of the dilution of their shares," you simply are not permitted to say, "Whoops, heh heh, can I just have those copies back?" You're toast; you're dead meat; you've had your head handed to you.

What's gone wrong here is time. Because we are geared towards heroic presentations, we keep our work under wraps until we go public with it (that is, publish it) at the big meeting. Until that moment, no one is allowed to look at it without our permission. It is secret.

But the Web is changing this. There's already software that lets groups work together on documents over intranets. And that capability is being built into the word processors themselves so that it'll be as easy to post a draft to a shared Web space as it is to send it to be sprayed on paper.

So, you'll be given an assignment and, just as before, you'll retire to your cubicle, but only for about half an hour. You'll write up some initial ideas, post them to the intranet—this feels like saving them into a shared folder—and you'll send out mail to the people you think can help you with this. (Here's how much attention you'll pay to where these people are located in the org chart: zero.) Your e-mail will say, and I quote:

> Old Man Withers wants me to solve the Parchesi problem in Tahiti. By next month! Yikes! So, I posted a couple of ideas at https://rsmythe.megacocorp.com/parchesi. I also put in some links to Donkeyballs' (oops, I mean Donnerby's) bogus report from last year, the one that didn't see the crisis coming. You can always count on Donkeyballs. ;-) There are also some links to a couple of sites I found when I did a search at the usual-suspect search sites.
>
> Let me know what you think. And remember, the doc I posted is just a bunch of BS. Kick it around, and let's get this thing going...
>
> Thx, guys and gals. You's the greatest!

This may not sound revolutionary, but consider:

- People used to keep their drafts secret for fear of looking like idiots, but now they post them and acknowledge they may be completely wrong.

- Work has gone from an individual task to a group task.

- The old model of keeping drafts secret until the moment of publication has been broken; ideas are now public from their inception.

- Ideas are assumed to be given out freely rather than hoarded.

- People are brought in not because they are in a chain of command but because they have necessary skills, share interests, and are fun to work with.

- Sober-sided reports that were the mark of professionalism are often replaced by humor-filled interchanges.

Where do secrets fit into this picture? Fear of letting information out would cripple this project; the report that would emerge would be far inferior to what arises from a free interchange of ideas.

Besides, the Web lets everyone talk to everyone, in every department, across divisions, with strategic customers and even competitors. There are no secrets.

Decisions, Decisions

Of course, you're not providing open access simply to fill people's heads with scurrilous thoughts and titillating tidbits. You want people to make better decisions. But open access to information also means that you've undercut your normal decision-making process.

Why do we have a decision chain in the first place? Ostensibly, it's because those up the org chart have a wider view as well as more experience. There is something to be said for experience, although it can thicken the skin as well as ennoble the mind. But if everyone has access to information, those on top no longer necessarily have the widest view. Being close to the customer and being in constant interaction with one's suppliers may bring an equally deep view into the business and its real possibilities.

Decisions are centralized also to enable accountability: praise for success, condemnation for failure. But every team member recognizes—and often resents—this fiction. I sat through an off-site meeting once at which the middle managers were handed Cross pens to reward them for their success (but really to buy loyalty to the man handing them out). Afterwards, one of the managers told me he felt dirty. Though his team had done the work, he got the pen. The pen was now a symbol of what he hated about his job. He would pass along the praise, of course, but clearly—he thought—senior management didn't appreciate how hard the team had worked. Being appreciated is not a commutative property—it requires eye contact, not the ritualistic passing of pens. And, of course, if the teams had failed, the senior executive accountable for the failure would have passed the criticism down and, not to be cynical, would likely find a way to dodge the bullet.

Does this mean that every decision will be collaborative? Of course not. But neither will every decision be taken by an individual.

We have a rich heritage on which to draw. Our culture has evolved many ways of making decisions simply because we have many ways of being together socially. For example, we seem to think having everyone vote works when it comes to running a country that can start wars, appropriate property, and execute malefactors, yet we assume it's a bad way to run a business. There are lots of reasons for governance through voting, including assuring that people have a say in setting policies that affect them, but one is particularly relevant to business: Wisdom is a property of groups. In most instances, groups are collectively smarter than their individual members and often make more sensible decisions. The fact that typically the only group in a corporation that gets to vote is the board of directors is not an accident; decision-making is usually more an exercise of power than an act of wisdom.

Of course, majority vote isn't the only way to make decisions. There's consensus, compromises, negotiations of every stripe, even counting eeny meeny. Yet for all this richness, in business we default to autocratic rulings. It seems a shame.

So, two outcomes are likely as the work of business increasingly moves online. First, we'll see more ways of deciding because we're seeing more ways of associating. Second, an important part of every project will be how you are going to decide.

Yes, this requires focusing on something we've often taken for granted before. But it will also open up for explicit discussion the nature of the social interaction in any particular project: Is this a group effort, a team with leaders, a mob action, a ventriloquist act, or some other type of human association? Even raising this for conversation in a group changes the dynamics, for it acknowledges the fact that there are lots of ways humans can work together—and every type of association is a matter of choice.

Unmanaging Rich Data

ALL THIS OPEN INFORMATION. SOUNDS LIKE A NIGHTMARE to most of us. But in fact, *information* is the wrong term for it; we just don't have anything better.

The term *information*, as we commonly use it today, is a product of the computer age. Before then, *information* meant something like *news*. The term took on special meaning first in information theory, where it received a mathematical definition (to the yawning indifference of the awaiting public) and then in the computer world when *data* was invented.

As everyone who's taken Computer Science 101 knows, information consists of significant correlations of data. "Ants #1–#100 died at 8:58" is data. "Ants #1–#100 ate mayonnaise from the office cafeteria at 8:51 and died at 8:58" is information.

Row and Column View of the World

Last_Name	First_Name	Start_Date	Employ_Level	SSN
Aardvar	Hyman	03-13-1992	J4	012-34-5678
Antear	Marjo	11-07-1998	B3	876-54-3210

"Information" is the stuff that goes into computers. And we all understand that to get the relevant facts about the world into our databases, we have to strip out a lot of the subtleties. For example, when we're populating our employee database, we have fields for "Name," "Start_Date," and "Salary," and maybe one for "Hobbies," but we certainly don't have fields for "Hates_Thai_Food," "Can't_Remember_Names," "Hums_While_Reading," and the things we know about our coworkers that together constitute a context for working with them.

We strip out the context because that enables us to manage information: We select rows based on the content of the columns, we sort and arrange the rows, we look for interesting correlations of rows and columns. In short, information is stuff we generate precisely to be managed with computers.

The Web isn't about information, however. While it takes a database administrator or data entry specialist to enter data into a database, it takes any idiot with a computer to post something—from naked pictures of your cat to an overheated manifesto—on an intranet or on the Web. And it's only going to get easier.

So, while we're populating our corporate databases with context-less, stripped-down information that can be managed, we're populating our new Web world with every type of artifact the human hand can devise without a thought about how it will be managed.

Information is built to be managed; the stuff on the Web is the product of the lack of management. Information is stripped down; the content of the Web is rich in its contextuality. These two sets of contrasts go together.

Rich Content and Human Voice

The stuff on the Web tends to be rich, not dry disquisitions loaded with charts and tables. Rather than a nicely printed report entitled, "An Analysis of Competitive Strengths and Weaknesses of Product #456-A," you're more likely to get "Why 'Gosh Honey You Smell Great for a Corpse'™ Sucks but Will Rule the Underworld Anyway."

There are bunches of reasons why this is so.

The Web is a document world. Eons ago, there was the Internet and it was populated by a subculture of Jolt-drinkin', four-eyed, research-crazed academi-geeks who used a Unixlike language to ferret out morsels of information. (Unix is the Klingon of cyberspace—an argot only the true fanatics learn.) Along came the Web with two simple additions to the Internet.

First, the Web replaced screens and terminal emulations with a much more familiar and useful way of presenting stuff to be read: documents.

Second, the Web made it easy to hyperlink to a document without requiring the author's consent. This made it possible to navigate the Web by clicking on content rather than by typing in path names.

The Web succeeded where the Internet failed, in other words, simply by adding a document front-end, and hyperlinking those documents together. The document user interface made it simple for people to get started with the Web. (Here's the instruction manual for a Web browser: If it's blue and underlined, click on it.)

This is important because documents are our most richly evolved type of data. Our culture has spent a couple of thousand years figuring out how to express virtually any type of thought on pages. Because we are so close to documents in all their forms, it can be hard to realize just how good we are at reading them and just how much contextual information they convey. We parse the structures of a page instantly and thus can tell the footer from the foot-notes, the header from the headlines, the byline from the lines of bile. Computers still can't match us at this; just ask any user of optical character recognition software.

The Web is a document-based medium. It is built to handle the richness of documents. And, interestingly, the very first improvements of HTML (the lan-guage in which Web pages are written) mainly concerned themselves with sim-ply enabling Web pages to look more like spiffy *printed* pages.

So, we're used to documents, documents are capable of handling a huge range of human expression and ways of structuring ideas, and the Web lets us maintain this sophisticated way of communicating.

The world of information on the Web is, therefore, a whole lot richer than the domain of database information in both content and structure.

But, wait, there's more!

The Web is a voiced world. The Web is the realm of the human voice. As we discussed in Chapter 2, your voice isn't simply the sounds that come out of your mouth. It's the way you present yourself in public through speech, writing, dress, body language, manners—virtually all that you do. The Web liberates voice by making it so damn easy to communicate and publish.

We have been trained throughout our business careers to suppress our individual voice and to sound like a "professional," that is, to sound like everyone else. This professional voice is distinctive. And weird. Taken out of context, it is as mannered as the ritualistic dialogue of the seventeenth-century French court.

The Power Table

Suppose you removed the table from your conference room and replaced the seats with armchairs. Suppose you turned it into a living room. How much would this affect your meetings? That's how much your meetings are about power, not communication.

We may be accustomed to the professional voice, but it isn't natural, God-given, or neutral: It's the voice of middle-aged white men who will do anything to keep people from seeing how frightened they are.

If you need to hear how the professional voice sounds, dig out any memo you wrote four years ago and compare it to how you'd write an e-mail about it now. A professional memo obeys implicit rules such as one page is best, no jokes, admit no weakness, spellcheck it carefully, and send it to as few people as possible.

Now, we write e-mails. They're short, pithy, funny, they sound like us, and we cc the CEO on a whim. That's why most of us don't want to use a word processor to write our e-mails. We want to be free of the expectation that we've spellchecked it or even re-read it before firing it off. We certainly don't want to waste our time monkeying with fonts and margins. At most, we'd like to be able to make words bold by hitting the keys harder.

E-mail enables us to construct our voices at our leisure, resulting in some odd artifices. A voice is, after all, a complex "thing." We have different voices for different environments and even for different people—we don't talk to our coworkers precisely the same way we speak to our children (well, unless we are very senior managers). Because most of our communications over the Web are "asynchronous"—i.e., not real time back-and-forth—we can construct our presence a bit more carefully. Our culture is currently in a phase where people are trying on voices, discovering what works and what doesn't work over e-mail, bumping up against the limits, and making lots of mistakes. For example, while e-mail can replace many meetings (primarily because at a physical meeting you can't skim over the remarks of dunderheads), e-mail is a profoundly bad medium for conveying personal criticism precisely because it is textual and thus not very *con*-textual.

> ## E-mail Expectations
>
>
>
> **W**hen you receive an e-mail you expect it to be:
>
> - Brief
> - Funny
> - Hastily written
> - Ill-considered
> - Thoughtless
> - Regrettable
>
> **All part of the charm!**
>
> —Christopher Locke

Here's another way the voice of e-mail is destroying committee meetings: After the carefully controlled meeting is over and the bigwigs are congratulating themselves on how well they managed it ("I think we got exactly what we needed out of that meeting, JB"), the "junior" people are back in their cubes firing off e-mails parodying the results and pillorying the personalities. Meeting go boom.

The return of voice is dooming not only the memo and the pointless, drone-a-thon meeting, it's also turning the corporate propaganda newsletter into a flat-out embarrassment. Instead, individuals' 'zines are popping up in organizations, written by people with points of view, human voices, and usually a sense of humor. For example, at Optika, a small software company in Colorado

Springs, Sean Spradling, a twenty-six-year-old member of the Marketing department just up and began publishing FORECAST THIS!, an internal 'zine that presents Sean's highly biased view of the market and Optika's marketing efforts. If "uplifting" characterizes most corporate newsletters, "skewering" characterizes FORECAST THIS! But its readers—the salesforce, marketing, and most of Optika—know to trust it, and look forward to getting it because it's written in a real voice stating the real truth. What a concept.

In a hyperlinked organization, voice plays the old role of the org chart, telling you whom you should work with. That Mary is the Under-VP of Expectation Deflations for the western semi-region tells you nothing. That Mary is wicked smart, totally frank, and a trip to work with tells you everything.

Thus do the formal bonds dissolve, replaced by the sound of the human spirit.

Telling Stories

The world is more like a huge set of messy hyperlinks than like a really big table of data. It is a world in which information isn't abstracted into some seemingly neutral means of expression but is always uttered by some particular human in that person's own voice.

So what happens to information management?

On the one hand, it continues much as it is. We still need databases that reduce people to numbers. Couldn't live without 'em. But we also should recognize that the increase in available information has made us feel stupider than ever. All the printouts, all the database dumps, and all the nicely formatted reports and spreadsheets with embedded charts are not describing our world to us. It's just not adding up. We have statistics but no understanding. And adding more and more information is only increasing the noise level.

We don't need more information. We don't need better information. We don't need automatically filtered and summarized information. We need understanding. We desperately want to understand what's going on in our business, in our markets. And understanding is not more or higher information.

If you want understanding, you have to reenter the human world of stories. If you don't have a story, you don't have understanding. From the first accidental wiener roast on a prehistoric savanna, we've understood things by telling stories. I don't mean fiction or stories heavy with plot; I mean narratives that string events together in time and show them unfolding.

For example, my young son in some sense understands World War II. His story is this: The Nazis attacked other countries and were winning until the U.S.A. stepped in and beat the Nazis.

A Russian child's story about World War II is likely to be very different: The Allies delayed opening a second front until the incredible sacrifices Russia made wore the Nazis down, and then the United States finally came in and finished the job.

The Knowledge Management Impulse

Knowledge management has become a hot topic precisely because we silently recognize that our information isn't yielding understanding. But information is unsatisfying precisely because it's managed; to make it manageable, we strip out context and voice. So, if we identify something called knowledge and then insist on managing it, we'll repeat the problem that gave rise to our desire for knowledge.

Conclusion? If you want to get past information, you have to give up the hope of managing your—and others'—understanding of the world. Also, you can't do it yourself: All understanding is social, by definition.

Both stories are ways of understanding the war.

My son doesn't understand the First World War because he doesn't have a similar sort of story, right or wrong. ("Once upon a time, there was an archduke....")

Here's another example. I worked at a company that tanked for lots of good reasons. When a bunch of us ex-employees get together, some of us say that it was because the product got too inbred and complex; others say that Marketing failed to predict the platforms the software would have to run on; others say that the management team was too focused on new products and ignored the bread and butter. None of us tell the same story. And that means that we, as a group, don't understand what happened.

That's a sign of trouble, as we point out in the previous chapter. The company's origins are part of its authentic identity. That identity gets expressed in stories that sound something like these:

- Our founders were living in a garage and came up with an idea for "mistake management." They thought it'd be great for law offices, but it turned out that lawyers are late adopters of technology. Then, at one of the law firms they called on, they noticed that the secretaries were mixing Wite-Out with cream soda because...

- We're 157 years old, and started out making faux papyrus, which was in great demand during the great Egyptian interior decorating craze in the mid-1850s. When that died out, we realized that we had manufacturing equipment that—if you just adjusted the thickness—could just as easily turn out prefab walls. And that headed us in the direction we're still in...

- This company was founded by tech weenies and got off a great product real early in the 3D fax market. That was a time when being first mattered more than anything, and the company lived on peanut butter and cheap drugs. But, you know how it is, you hit a wall where you have to bring in the suits. So, we went out and hired the former COO of Sears...

When you get past the mission statement and the slide showing why your current market share and revenues are making Croesus envious, and you start to tell your story, only then do people begin to understand your company.

And it's not just companies that have stories. Every sale worth knowing about has one ("It looked like the bad guys were going to win this one, so I wrote this e-mail, see, and sent it to this guy I know..."). Every repair job has one ("I tried everything in the book to get the X405 to work, including repacking the bearings, which is a total pain. And then while I was tightening the booster ring, I noticed the damndest thing..."). Every product has one ("We couldn't figure out why no one was using the cup holders in the Deluxe model, so we did a study and we discovered that the engine is so powerful that people were afraid to let go of the wheel. So we decreased it from 36 to 12 cylinders and scored a hit with the scaredy-cat driver market...").

We live in stories. We breathe stories. Most of our best conversations are about stories. Stories are a big step sidewise and up from information:

- Unlike information, they have a start and a finish.
  The order counts a lot.

- They talk about events, not conditions.

- They imply a deep relationship among the events, a relationship
  characterized overall as "unfolding" as if the end were present
  in the beginning—as of course it almost always is (as was fore-
  told, in a fractally recursive sense, by Aristotle at our culture's
  beginning).

- Stories are about particular humans; no substitutions allowed.

- Unlike a set of economic forecasts or trends analysis, they do not
  pretend to offer the certainty that life will continue to work this
  way. (On the other hand, the story is more likely to be correct
  than the forecast because it takes all of our current understanding
  of the world to accept a story.)

- Stories are told in a human voice. It matters who's telling it.

So, stories are not a lot like information. But they are the way we understand.

How to apply this to your workaday world? You already have. When you are telling someone how you won this account or lost that one, when you are explaining why the competitor's trade-show booth was a disaster, or when you are telling a financial analyst how the market got to be as wacky as it is, you're already telling stories. You can't help it. You're human. Stories are how we make sense of things.

Anything else is just information.

Seven Ways to Tell Stories

1. Ban the opening joke. Begin your next PowerPoint presentation by saying, "Let me tell you a story..." and then recount what made the market the way it is, what got your company to come up with such an incredible product, and what obstacles particular customers faced and overcame by using your product.

2. Make sure the forms you use to "collect knowledge" have big empty boxes in them so the story can be told.

3. Every meeting with a potential partner, every exciting sales meeting, every important encounter with customers can best be told as a story. Do so.

4. Turn your next white paper into a narrative.

5. Collect the stories of your business and publish them on an intranet site.

6. Reward the tellers of good stories. They're the people everyone's listening to anyway.

7. Rewrite your mission statement as a corporate story. In fact, wouldn't a narrative version of an annual report help the company more than the usual hearty prose and canned snaps of happy employees?

Brokenness

STORIES ARE A WAY to understand a world that can surprise us. But in Fort Business, surprises are a sign of the failure of management. Management aims at predictability and it tries to get there via control.

The urge to manage is deep in our culture. It ultimately is defeated by the fact of human fallibility.

It's in the Web's nature to "always be a little bit broken" because it's decentralized. No one is in charge of making sure that the page you're trying to get to hasn't been taken down. There's no one to fix the Web, no one to plan it, and no one to complain to.

In fact, all big systems are broken. We don't always see that because what counts as broken is a matter of perspective. For example, on the phone system sometimes we get busy signals, and sometimes the phone rings and rings and no one answers, but we choose not to count those as signs of brokenness. If the telephone system chose to treat busy and unanswered phones as broken, it could make answering machines a standard telephone service. We could even complain that we have to memorize long strings of numbers, instead of having cute phone "numbers" like david.weinberger.the.balding.one@brookline.ma.usa.

We choose to see the phone system as basically not broken, and choose to see the Web as inevitably broken. Why? Because fallibility is an endearing trait that seems to be a requirement for community. We of course want the people we work with to do everything they can to meet their commitments to us, but we also may find it hard to trust people who refuse to admit fallibility—their own and others'. We are intensely uncomfortable with people who have no weaknesses. For example: Michael Jordan, Jesus, and my older cousin Don.

The Web's frailty makes it more human, less threatening. It also lets us move faster. For example, Mark Gransee, VP of Information Systems at Eddie Bauer, said (in an article in INFORMATIONWEEK):

> In the old cycle, you could...hit analysis paralysis. Now you can't
> be afraid to make a decision just because the conditions are going
> to change and make that decision obsolete.

He adds that perfectionism isn't allowed: "You just have to do the best you can."

Meanwhile, at Owens Corning, Mike Radcliff, CIO, said (also in INFORMATIONWEEK):

> Our staff has to be able to work with incredible ambiguity, be
> self-confident, simplify and trust others.... Most of all we have to
> embrace "good enough" reengineering, good enough that we can
> progress...not necessarily what we'd do in the ideal world.

But it's not just systems that are imperfect. More important, so are we humans. Say it with me: Humans are imperfect. I am imperfect.

Feels good, doesn't it?

We often use the phrase "knowledge is power" to make it seem that hierarchically granted power is justifiable. In most hierarchies, however, knowledge

Cluetrain's Own Politics of Rightness

Cluetrain itself is prey to one particular Being Right stratagem: Everything you know is wrong (in the words of Firesign Theater). Thus, our first temptation is always to find the way in which current beliefs are mistaken, preferably in some very clever and deep way.

Knowing this, however, has not helped us get past it. Many more years of therapy are called for.

isn't power, it's a weapon. Being right advances you and being wrong is a defeat. That sucks.

You can see the politics of "being right" throughout most organizations. People win arguments—and thus secure their position in the hierarchy—through the cutting remark, through megatonnage of evidence, through agreeing with industry consultants, and through the smug refusal to ever admit being wrong.

But wrongness has a lot going for it beyond the fact that some things can only be learned through trial and error. For example:

- Some people are great at generating ideas but terrible at thinking through their impact. You want them to have as many bad ideas as possible because they will thereby randomly generate more good ideas. (I tell my clients that I try to maintain a 9:1 ratio of bad ideas to good. And, no, I can't tell which are which. If only.)

- Errors are how assumptions become visible. And there is nothing more valuable than a newly discovered assumption, because only then can you see what's holding you back and what could propel you forward.

- There's too much to know, so all important decisions are, to some extent, random. By being free to make errors, you can try more paths until you stumble on one that takes you somewhere interesting (albeit probably not where you at first thought—mistakenly —you should be heading).

- Errors remind us that we're fallible humans. A company that's too embarrassed to admit mistakes and that builds a culture where

> being wrong is humiliating literally is denying what it is to be
> human. And you will pay the price—in this world, if not in
> the next.

- Mistakes give us something to talk about.

- Being wrong is a lot funnier than being right. The right type
  of laughter—laughter at what the mistake reveals about our
  situation rather than laughter aimed at a person who dares to
  be human—is enormously liberating. In fact, laughter is the
  sound that knowledge makes when it's born.

Does your company have "zero tolerance" for error? Can you change your
mind without losing status? If so, consider engaging in the radical politics of
wrongness. Go out and commit a whopper. Then embrace it publicly.

It's a good feeling. It's liberating. It's how you find your voice.

Blurry Boundaries

WEBS HAVE BLURRY BOUNDARIES. FORT BUSINESS, ON
the other hand, makes an enormous investment in maintaining the integrity
of the walls.

Hyperlinked organizations never met a wall they liked.

In the world of closed rooms and weekly meetings, you're a member or
not. To join, you have to commit to sitting in a room at a particular time. In the
open, hyperlinked world, it requires nothing but a few clicks to check out what
a particular group is doing. You join their e-mail discussion group or visit their
group intranet site. Zero commitment. So membership isn't a yes-or-no decision. You can browse with all the lack of commitment the word implies.

When the hurdles to membership lower, the boundaries blur. The blurring
isn't occurring only inside of the Fort. Businesses are building extranets to
enable their strategic partners to access information. There are hundreds of
examples of this, in industries that range from retailing to drilling for oil to
distributing T-shirts to the people who print slogans on them.

In many cases, extranets are used to get the paper out of the system. This enables process automation and cost savings, which are good things. But some companies—and someday, all companies—are going farther than that, giving their partners and customers access to their own intranet, so they can see the sausage being made.

Intranet technology is sophisticated enough to let you control exactly who has access to what, so it's no longer an all-or-nothing proposition. You can let customers see product-design discussions but keep them from seeing what its competitors are saying to you; you can let a supplier check the processing of a payment but keep it out of the pages where your accountants are evaluating bids. You have all the flexibility you need. The old excuses for pulling up the drawbridge and keeping everyone out entirely just don't hold.

Why not let your customers see your product-design process? They know that it's not perfect. They know you're going to go down wrong paths, you're going to abandon pieces you thought were locked in, you're going to squabble senselessly over trivia. That's what business is like.

Every business is dysfunctional because everything human is at least a little bit broken. It's not an accident. It's the human condition.

So what are you protecting your customers from? The obvious truth they know and live with every day? Just exactly whom do we think we're fooling?

Companies that let their customers and suppliers into the process early on deliver better products. And they forge the bonds of trust and delight that are the only ones that work in the "frictionless" Web.

But maybe you need more than the promise of riches. Perhaps you need the fear of failure to motivate you. So, here it comes: Suppose you use your extranet solely as a secure publishing site or for automating transactions that otherwise require paper, rubber stamps, and file folders. This will decrease your expenses and your time to market. Excellent. But if that's all you do, the first companies that knock down the walls to their customers and suppliers will eat your lunch and then beat up your children for their lunch money.

Imagine the Foobar Company, the leading supplier of pen chains to the banking industry. Its development process calls on it to come up with a marketing requirements document that results in a product spec that in turn results in a new product. The entire development process is done behind walls because Foobar can't let its competitor, Wumba Chains, find out what it's doing. But now Foobar has discovered that Wumba is letting its customers into Wumba's product-development processes way early in the game. As a result, Wumba's customers are ready with purchase orders the day the product ships, whereas Foobar's customers need months of explanations and wooing from the sales force. And while Wumba's customers feel they're getting the real poop, Foobar's customers find the carefully constructed and controlled press releases and product brochures to be barriers more than helpers. They have to sort through them to try to get a sense of what's real and what's wishful thinking.

So, Foobar decides to open the floodgates. Customers and suppliers are poking all around the innards of Foobar Company. These "outside" companies are seeing the actual workings of the company, and that means they are getting to know the individuals in the organization. They're learning that for questions about the safety features of pen chains, they should listen to what Paolo has to say, for information about retrofitting cars for advanced pen chains there's no better source in the industry than Mary, and when it comes to addressing future ideas for pen chains you should never, ever, pay attention to what Amit says.

As this sort of knowledge gets absorbed, the "outsiders" start dealing directly with the individuals and hyperlinked groups in the organization. If a partner needs to know how the pen chains are going to work with the new government regulations, why go through Legal or Regulatory Compliance or Marketing, which will take six months to formulate some ass-covering BS, when you can pop into a work site or have an e-mail discussion among the people who really know what's going on and will tell you the truth?

And when you pop into this group, are you going to know or care that some of the members are in fact other partners of Foobar? If Juan is an articulate, knowledgeable, trustworthy voice in the discussion, will you know or care that he is in fact a supplier or an industry expert who works for one of Foobar's customers?

To the outside, the company begins to look like a set of hyperlinked clusters who select themselves based on trust and respect and even their sense of fun. The trust is built through the quality of voice of the participants: That is all that counts in a hyperlinked team.

The business now consists of a shifting set of hyperlinked groups, self-organizing, inviting in participants based on the quality of their voice, regardless of where—and whether—they are on the org chart. Management is simply an impediment to these groups. In fact, rather than employees feeling that they must constantly justify themselves to management, management now needs to give workers a single reason why it should be involved in the life of the business it used to believe it ran.

Hyperlinks subvert hierarchy. Hyperlinks subvert Fort Business.

Business is a conversation.

The Economy of Voice

NO ONE'S ASKING YOU TO DECIDE IF YOU WANT TO RUN your business using the Web. It's a done deal. The Internet has already set expectations for how connections ought to work. The gulf is there; a gulf caused, ironically, by the abundance of connection.

The Web is the sum of these connections. It isn't a medium, a new type of intercom, or an invention like really cool wristwatch walkie-talkies. It is a broad, open place that lets everyone touch everyone else and touch every digit of information by twitching a wrist and tapping a single finger.

What connects you to me to everyone else are Web pages and e-mail and chat and discussions. These are all artifacts of human voice. Each is deliberately created and put forward as our public self, the self that is closest to us and, paradoxically, least knowable to us.

An economy of voice. Has there been such a thing since the Athenians talked democracy into existence?

The voices are heard in conversations. That's why the Web has its transforming power: It turns out the fundamental elements of our world have been products of deep conversations all along—conversations carried on by philosophers, artists, poets, and other crafters of language. Had those conversations across the generations been different, we would not have the world we do.

These particular conversations have given rise to a deterministic, causal world in which outputs result from inputs according to natural principles and self-evident rules. The world's mechanism depends not only on predictable—and thus interchangeable—parts but also on the centrality and predictability of laws of nature, principles of behavior, and time itself, a recent achievement in our history. (Socrates never said, "Hey, Alcibiades, what do you say we meet at the corner of Hesiod and Pericles at three-fifteen? Later, babe.")

Physical laws, rules of behavior, contracts, schedules, deadlines, professionalism, org charts, and management practices are all types of connections. They all are attempts to control not only the object of the connection but also the nature of the connection itself. Why? Because they promise control over the two things we fear most: the vicissitudes of our world and the passion of our selves. As a manager armed with a theory and the latest business book, I not only know what to do, I know who to be.

Then the Web crept into our offices under false pretenses. We thought first it was a library of information. Then we thought it was a publishing medium. Then we thought it was a toy or a dangerous distraction. But in fact it is a conversation of a new type, free of the need to get permission from Dad and his army buddies.

New types of connections. The heart flowing to other hearts. A new rhythm. A new causality. A new understanding of power. Conversation that understands that it isn't a distraction from work, it's the real work of business.

The Web is hitting business with the force of a whirlwind because it is a whirlwind. The closely held, tightly packed, beautifully tooled pieces are being pulled apart. They are rebinding themselves in patterns determined by the conversations that are occurring in every conceivable tone of voice.

The character of business is becoming the same as the character of the Web—an explosion reconfigured by the intersection of hearts.

EZ Answers

CHRISTOPHER LOCKE AND DAVID WEINBERGER

Tell 'Em What You Told 'Em

FROM THE FURTHEST HISTORICAL REACHES OF JUMP STREET, markets have been conversations. Craft and voice were joined at the hip—what you made was how you spoke. But then it turned out that the world was round, there were other places across that Big Blue Wet Thing, and trade routes got longer, natch. Producers became further removed from markets. Gradually, marketing became an abstract pipe down which producers shipped products to customers, though nobody would invent FedEx for several centuries. Somewhere along the line, speech and craft lost each other's phone numbers.

At the beginning of the Industrial Revolution, new power sources replaced much human grunt work. Producers immediately saw that this was a Good Thing. Moreover, they saw that repeatable processes and interchangeable parts were an Even Better Thing, as such mechanization led to significant economies of scale— a fancy way of saying more money.

By the time the twentieth century rolled around, industry hit upon an even more potent multiplier: interchangeable workers. The assembly line turned workers into machines. Through this stroke of genius, craft skill was effectively hosed, and workers were told to shut up and do what they were told.

Economies of scale also required economies of management. Telling them what to do efficiently required a new form of business organization. Bureaucracy fine-tuned the division of labor needed to make this new setup work, and a breakthrough concept called the org chart determined who got to speak at all. Welcome to management by command and control. This resulted in *huge* economies of scale—a fancy way of saying "robber barons."

Mass production led to mass marketing, which led to (ta-da!) mass media. Broadcast applied the fundamental mass-production brainstorm to marketing communications. This development signaled the dawn of junk mail. Corporate speech became mass produced "messages" jammed into a one-way spam cannon aimed at a dream that hasn't faded since: interchangeable consumers.

Ignoring the clear lessons of history (for example, the nuking of Hiroshima and the saturation bombing of Dresden), upstart "foreign" companies started selling into markets the United States figured it had permanent dibs on. Guess again. The global economy threw a monkey wrench into the sweet deal that was mass production. Established markets broke up into a zillion micromarkets, leading to an explosion of new products and services: Now you could get a car specifically designed for your urban, sports, just-divorced, hockey-fan lifestyle. Or whatever.

New knowledge was desperately needed to fuel this expansion, and this is when companies discovered what workers had long suspected but never talked about except in the washroom: Management didn't know its ass from a hole in the ground. (See clear lessons of history, above.) While managers had gotten really good at bossing people around, they didn't know much about how things actually got made. This naturally resulted in many exciting high-level executive-type conferences about "The Knowledge Deficit."

Slowly (some are still attending summer school), companies began to realize that workers knew more than they'd been letting on—mostly because no one had asked them for about a hundred years. This led to the reemergence of craft in the workplace, and a concomitant revaluing of speech—a fancy way of saying "lead, follow, or get out of the way."

Ideas, talk, and conversation were now encouraged among workers because they helped to deliver what organizations so desperately needed: a clue. During this period (which unfortunately ended in large measure due to "downsizing"—but that's another story), "empowerment" became the watchword of the day, and org charts were upended or tossed out altogether at companies like GE, Ford, Motorola, Corning, Cadillac, and Federal Express (which by now, of course, had been invented). This was the era of the U.S. Department of Commerce's Baldrige Award when the pursuit of Quality—always capitalized—took on a decidedly religious fervor. What Quality really meant was: "We changed our minds. Please don't check your brain at the door."

While speech was actively elicited from workers because it carried suddenly invaluable knowledge, it was not yet sought in any significant way from customers—a concept still perceived by many corporations as more dangerous than godless communism and universal healthcare combined. However, due to market fractionation, "consumers" had already become far less interchangeable.

Then along came the Internet and all hell broke loose.

Just as the global economy had precipitated exponential growth in the array of choice among new products and services, so the Net caused an explosive proliferation of choice among new information sources. The broadcast model faltered and failed online. Embarrassing attempts to force it to work, such as "push," were quickly swept under the rug—where, in the form of a large pig-in-the-python lump, they continue to trip up wannabe online businesses.

By its nature, Internet technology encourages open distributed speech, a fancy way of saying "tellin' it like it is." The human voice is the primary attractor, both *to* the medium and within it. Markets and workers are once again crafting their own conversations, and these conversations are also about craft—things we do that we actually care about.

As a result of the profound and unexpected changes wrought by the Net, the two-hundred-year-long industrial interruption of the human conversation is finally coming to an end, both inside companies and in the marketplace. That's what www.cluetrain.com basically had to say when it hit the Web in 1999.

And you should see the flame-mail we got! This is because we're now living in a period that could be called "the Hangover." Command and control is widely perceived as dysfunctional, but it's a hard-to-break habit. Many business leaders are well aware that bureaucratic hierarchy works against needed knowledge and communication, yet inertia is a powerful force. ("The cluelessness is strong in this one, Darth.")

Though the Internet represents an unprecedented invitation to break out of this impasse, many organizations today resemble the Berlin Wall—monoliths interposing themselves between the internal conversation of the workforce and the external conversation of the marketplace. They are still pumping out mass-produced messages, still trying to control workers and consumers, still trying to create mass markets based on old industrial models.

Via intranets, workers are already speaking among themselves. Via the Internet, markets are already speaking among themselves. The convergence of these two conversations is not only necessary, but inevitable. Why? Because markets, unencumbered by corporate bureaucracy and the need to ask permission at every turn, are learning faster than organizations. Markets are therefore coming into a new ascendancy, a fancy way of saying, "We rule, dude!" And increasingly, we value only two qualities:

1.   The engagement and passion-for-quality of genuine craft.

2.   Conversations among recognizably human voices.

The simple, if painful, prognosis: organizations must encourage and engage in genuine conversation with workers and markets—or go belly up.

So what, if anything, can businesses do at this juncture? They can begin by searching out people within the organization who understand what's going on. In almost every case, they're there. Make friends with them. Make friends with the marketplace again. Start listening. Find your voice. Then start talking as if your life depended on it. It does.

Business is being transformed, but not by technology. The Web is simply liberating an atavistic human desire, the longing for connection through talk. That's the one constant throughout our evolution, from caves to mud huts to open-air bazaars, from city-states to empires, nations, interdependent global powers. We've always conversed, connecting to the people of our world in our authentic voices. We connect to ourselves the same way; that's the mystery of voice.

But part of us still has a deep resistance to the unmanageability of the Web. We keep wanting to contain it within a business model, to build it into our business plans and see it as yet another "opportunity" for more/cheaper/ faster/better business-as-usual. E-commerce, oh boy! *Ka-ching!* The sound of the cash register is all too often the sound of attempts to co-opt the Web. To tame it, domesticate it, make it more familiar. To shoot it, stuff it, and mount it in the corporate board room along with the other trophies of corporate conquest.

At the same time that it spooks us, we're fascinated. We've been waiting for the Web to happen all along. We've been hating our jobs for generations. We've been longing to speak in our own voices since we made the Faustian deal to keep quiet in the first place. The Web is not aimed at business in particular. It wasn't built for business, it isn't fundamentally about business, and it can't be controlled by business—any more than the Internet could be controlled by the Pentagon that sponsored it.

The Web is inherently and intrinsically free. Businesses will perceive this fact as either a blessing or a curse depending on how much they value free-dom, a quality of mind and heart not typically underscored in the average cor-porate mission statement.

Loaded Questions

Literature is the question minus the answer.

Roland Barthes

If love is the answer, could you please rephrase the question?

Lily Tomlin

OVERHEARD AT THE COCKTAIL PARTY THAT WAS THE 1990s:

"So what's going to happen with all this Web stuff?"

"Where's it going? What are the trends? What are the directions?"

"Who's going to win?"

And they say there's no such thing as a dumb question.

Continents drifting across the oceans have trends. Bullets have directions. Cannonballs have trajectories. The future doesn't. The future is the intersection of choice and interruptions. The Web—what a surprise!—is more like the future than a cannonball. It will be what we make of it.

This leads to a funny conclusion. Ironic, actually. We ask questions about the future of the Web because we think there's a present direction that can be traced into the future. But in fact, the questions we ask aren't going to predict the future. They will create the future.

Not to get all heavy and ontological, but since questions are a type of conversation, it looks a bit like conversations give the world its shape, doesn't it? Questions do the spade work. They prepare the ground for answers. Be careful what you ask or you just might become it.

So far, the questions about the Web we hear the most—the ones some journalists use to stir up fear, the ones some politicians use to prepare us for the revelation that they just happen to be our saviors, the ones most businesses use to keep us in line and to sell us more stuff—these questions are actually decisions to look for the same old things. Their assumptions are wrong, and the answers they call for are mean-spirited. The questions themselves are intended to confuse the issue, and the answers are nothing but the smirk on the face of someone who just proved himself right.

There are other questions possible, better questions. Questions that come from the heart, not the wallet, the gonads, or the lobe of the brain responsible for smugness. Questions that open the future instead of making sure the dead bolt on the door is nice and tight.

For example, take the ever-popular question, *Will the Web become a broadcast medium? Will it become TV?* That's vitally interesting to media titans who see the Web as a threat to how they make money. But that's not a question of the heart.

What the heart wants to know is, When the buttons at our fingers let us talk with the polyglot world's artists, how will we cope? What will we share as a culture and community? What will we talk about together? What will we laugh about? What type of laughter—mocking, ironic, cynical, sinister, belly-shaking guffaws—are we going to hear? Will we find we all share a common sense of humor, or will we learn to laugh in new languages? When will we record the first case of Web inebriation, a trans-global xenophilia induced by pure, uncut connectedness?

Here's a question beloved of industry analysts and others who think the point of conversation is to appear smart: *How quickly will commerce move to the Web?* Let's trot out the charts and studies, confident that at least one of them is going to turn out to be right.

But is this question really so important, or does it just address a detail about timing? Is your business going to be transformed if it turns out we're not going to hit the gazillion mark until 2004 instead of 2003?

But there is a heartfelt question lurking here. It has to do with the things of the world that quench our thirsts and raise our souls. It has to do with our fear of replacing the shops—and the neighborhoods they enable—with a paper-souled efficiency that lets us search out and consume commodity products at disquietingly low prices. We're afraid that the last shred of human skin left on the bones of commerce is about to come off in our hands. We want to know how we'll reconnect to the other people in the market: buyers and sellers, people we know or whose faces are the landscape of our life in the agora. And we have this fear precisely because the e-commerce question has been asked wrongly so often, as if once commerce becomes virtual it will become cruelly automatic. We need to ask the heartfelt question about how we're going to talk about the things we care about, or e-commerce will indeed become nothing but the soundless scrape of coins over the wire.

Here's another question—top of the hit parade, actually—that steers us in a wrong direction as surely as asking, How can I drive straight to Hell, buddy? The question is, *What are we going to do about pornography on the Web?* This question seems to have nothing to do with business, but in fact it goes straight to the heart of maintaining a corporate wall between employees and customers, between internal and external behavior.

The question has to do with drawing the line between the public and the private—no trivial matter since looking at the line is the quick and dirty way we decide who's civilized and who's savage, who's refined and who's a brute. The line between public and private is, of course, arbitrary, although we adamantly deny this by using every method of intimidation, from the law to dirty looks, to maintain it. The main point of the line is that there be a line— one that we can control.

And then we turn on our computer and filth comes pouring out of every orifice, from our e-mail inbox to our browser. Go to whitehouse.com and you discover it's a porn site. (Hint: next time, try whitehouse.gov.) Open an e-mail titled "The info you asked about..." and get lewd invitations. Mistype a single letter of a Web address and you're staring at strange genitals in strange configurations. The Web isn't just redrawing the line, it's changing the nature of the line, making it explicitly permeable. But a new type of line means a new type of public.

So, our hearts ask questions, with dread as well as excited curiosity, about the new public world and its relation to the private. What is the relation of our night selves and our day selves, our self behind the company walls and outside of them? Why do we think of our private selves as our real selves? What would privacy be like if it weren't connected to shame? What is the fierce price we pay for every desire, every whim, every idea we stamp "Secret"? To what degree are shame and embarrassment the expression of the will to control? If we abandon the illusion of controlling private behavior, what type of public-ness will we have? How is the control we yearn to exert over the behavior of others—at work and beyond—identical with the white-knuckle control we need to preserve our selves?

More questions meant to distract us: *How will we know what's junk on the Web and what's worth believing? How will we avoid being fooled by anyone with a plausible story and a Web address? What will be the new criteria, the new marks of authenticity?* These questions express a longing for someone to take charge of our knowledge. We want experts and authorities, just as we crave censors more than we crave sex and prefer certainty to freedom.

But our hearts have a different set of questions: When we can't rely on a central authority—the government, the newspaper, the experts in the witness box—for our information, what new ways of believing will we find? How will we be smart in a world where it's easier to look something up than to know it? How will we learn to listen to ideas in context, to information inextricably tied to the voice that's uttering it? How can we reverse our habit of understanding matters by jumping to further levels of abstraction and instead learn to dig into the concrete, the personal, and the unique, told as stories worthy of our time?

WE ARE—ALL OF US—ASKED QUESTIONS LIKE, HOW WILL we manage (control) virtual workers in a distributed organization? when our hearts want to know how we are going to live with our families again.

We are asked, How are we going to keep our children safe on the Internet? when our hearts also want to know what it would be like to be a child who can talk within the world's society of children.

We're asked, How can you tell if the person you're talking with is really the person you're talking with? when our hearts want to know what people we will really become online and what having a disembodied identity will mean.

We are asked, How are the poor people of the world going to get Internet access? when our hearts want to know how we can connect with the poor of the world, because there isn't a single person we don't want to talk with. And once we talk, we know the conversation will make palpable the injustice of today's economics.

Our job now is not to answer questions. It is to listen past the questions based on fear and to hear the questions of the heart. Why? Because the proper answer to a heartfelt question is a conversation, and conversations make the world.

Hit One Outta the Park

GOOD THEN. THAT OUGHT TO PUT ALL THE WRONG-headed questions to rest, right? No, of course not. Business-as-usual being what it is, the questions never quit. Companies have said yes, the Cluetrain ideas are interesting, but give us a place to start. A methodology. A suite of best practices maybe. A set of guidelines. For God's sake, something!

"What's the bottom line?" they want to know. "How can my company profit from the coming transcultural train wreck? How can we leap tall buildings in Internet time, innovate faster than a speeding data packet, and establish Peace, Justice, and the American Way in hyperlinked global markets?"

Well…OK. Because you've been so patient and read so bloody much, we'll let you in on the Secret of Our Success. Just follow the twelve easy steps below and you're sure to be on your way to fame and fortune in the exciting new world of Webusiness. (Caution: It is vital that you follow these steps precisely in the order given. Otherwise, we are not responsible for the mutant hellspawn you may inadvertently call forth from the realm of the undead.)

The Cluetrain Hit-One-Outta-the-Park Twelve-Step Program for Internet Business Success

1. Relax

2. Have a sense of humor

3. Find your voice and use it

4. Tell the truth

5. Don't panic

6. Enjoy yourself

7. Be brave

8. Be curious

9. Play more

10. Dream always

11. Listen up

12. Rap on

Do these things and you just can't miss.

Of course, there's as much distance between this advice and the decisions you make every day as there is between "Go forth and multiply" and "100 Ways to Pick Up Hot Chicks and Radical Dudes." Still, we yearn for easy advice. It's so hard to give up the old wish for stimulus-response marketing and management. Hard to go back to the days of the "talking cure" when psychotherapy meant years of slogging through memories and dreams instead of a slap on the back, and instructions to "nurture the inner child" and eat two bran muffins every day. Hard to forget the televised version of ANNA KARENINA that goes from start to finish in two hours (the train comes to a screeching halt just in time) and reopen the musty volume and soak into every snow-flecked page.

Look, we'd love to derive twelve happy instructions from the wash of ideas swirling around us. Really. We could market those puppies like Tang in a sauna. Seminars, workbooks, T-shirts, coffee mugs...

But it doesn't work that way. This is an existential moment. It's characterized by uncertainty, the dissolving of the normal ways of settling uncertainties, the evaporation of the memory of what certainty was once like. In times like this, we all have an impulse to find something stable and cling to it, but then we'd miss the moment entirely. There isn't a list of things you can do to work the whirlwind. The desire to have such a list betrays the moment.

There may not be twelve or five or twenty things you can do, but there are ten thousand. The trick is, you have to figure out what they are. They have to come from you. They have to be your words, your moves, your authentic voice.

The Web got built by people who chose to build it. The lesson is: Don't wait for someone to show you how. Learn from your spontaneous mistakes, not from safe prescriptions and cautiously analyzed procedures. Don't try to keep people from going wrong by repeating the mantra of how to get it right.

Getting it right isn't enough any more. There's no invention in it. There's no voice.

Maybe we'd have more luck with the Cluetrain List of Don'ts than with a List of Dos. The first ninety-four items would be things like: Don't snoop on your employees, don't build knowledge management systems and corporate portals that are nothing but funnels for the same old propaganda. Don't hire people who claim to be experts at increasing morale. And right at the bottom of the list, number ninety-five, would be the most important one: Don't rely on lists, self-styled "gurus," or business books.

Scary, isn't it? Good. You ought to be scared. That's a realistic reaction. You want comfort? Invent your own. Exhilaration and joy are also in order. But face the facts: The tracks end at the edge of the jungle.

CHAPTER **[7]**

Post-Apocalypso

CHRISTOPHER LOCKE

We will strive to listen in new ways—to the voices of quiet anguish, to voices that speak without words, the voices of the heart, to the injured voices, and the anxious voices, and the voices that have despaired of being heard.

Richard M. Nixon, first inaugural address, 1969

IRONY IS PERHAPS THE MOST COMMON MODE OF INTERNET communications. The Net didn't create the mentality, but it did come along just in time to give it new expression. Nixon speaking about unheard voices of the heart from the height of the 1960s is a prime example of why most people have despaired of ever being heard at all. And of why they've stopped listening for answers from above—from Big Government, Big Business, Big Education, Big Media, Big Religion. With few exceptions, the interlocking agendas of these monolithic powers have become utterly divorced from the constituencies they were originally conceived to serve, their interests as remote from our daily lives as the court of King George was to the American colonies in 1776. And you know what happened then.

So are we calling for a revolution? What would be the point? The only revolution that matters is already well underway. And by the way, since it's not being covered by CNN and Fox, we're winning.

You say you didn't notice anything out of the ordinary? Nor were you supposed to. Invisibility and ignorance are powerful weapons.

Ignorance is not a value you often hear extolled. Let's make up for lost time. Here's how it works; it's pretty simple. When you ignore people long enough, they begin to feel invisible. Because your important concerns do not concern them, they begin to figure it's a two-way street. They begin to ignore you back. Pretty soon they're thinking Al Gore is some hockey player from Winnipeg, and Warren Buffet...isn't he the guy who does late-nite infomercials for cut-rate country western CDs? Three easy payments and it's yours? Yeah, but who really cares.

Ignorance is power. A maxim often heard online is that the Internet routes around obstacles, meaning it ignores them. In its early phase, the Net ignored business; Internet audiences simply weren't interested. And the feeling was mutual. Business ignored the Net for a long time, not seeing it as what it thought a media market should look like, which is to say television. This mutual ignorance served as the incubator for a global revolution that today threatens the foundations of business-as-usual.

Before any Old Order of Things can be given the final heave-ho coup de grâce, it's necessary to create a parallel infrastructure controlled by people acting in cooperation for their own benefit and mutual support. One thing any such effort requires is an extraordinarily efficient means of communication. We didn't used to have one. Telephones just didn't cut it.

Then, irony of ironies, along comes the Internet. It was as if we'd ordered it from Amazon: "Hello, U.S. Federal Government? Yes, we'd like one totally open, high-speed data backbone. Uh-huh, and charge that to the Department of Defense, why don't you? What's that? What do we want it for? Oh, just chatting about stuff. You know, this and that..."

Invisibility is freedom. At first it feels awful that no one can see you, that nobody's paying attention. Darn! But you get used to it. We've had two hundred years to get used to it. Then one day you find yourself on a network, networking, and it dawns on you that it's like walking through walls. Wow! Like some comic-book-mystic Ninja warrior! That's pretty cool. You can get away with saying things you could never say if anyone took you seriously. Dilbert is just a comic strip. Hah-hah. Beavis and Butthead is just a cartoon. Heh-heh. And if anyone comes sniffing around and wonders if this Internet stuff could be maybe dangerous, culturally subversive, it's oh, hey, never mind us. We're just goofing off over here on the Web. No threat. Carry on. As you were.

But we aren't just goofing off. We're organizing: building and extending the Net itself, crafting tools and communities, new ways of speaking, new ways of working, new ways of having fun. And all this is happening, has happened so far, without rules and laws, without managers and managed. It's self-organizing. People by the millions are discovering how to negotiate, cooperate, collaborate— to create, to explore, to enjoy themselves.

But what's the point, asks business? Business always wants there to be a point, a goal, an objective, a plan. Otherwise, how would we manage?

There never was any grand plan on the Internet, and there isn't one today. The Net is just the Net. But it *has* provided an extraordinarily efficient means of communication to people so long ignored, so long invisible, that they're only now figuring out what to do with it. Funny thing: Lawless, planless, management-free, they're figuring out what to do with the Internet much faster than government agencies, academic institutions, media conglomerates, and Fortune-class corporations.

So what is the Net really good for? Besides chatting, that is. Well, there's the small matter of coordinating distribution. Remember those ancient markets from way back in the first chapter where we talked about trade routes and the cities that grew up where they intersected? Where caravans arrived with exotic merchandise and tried to sell their wares:

"Figs here! Delicious figs!"

"Give me one. Figs want to be free."

"No way."

"I won't buy from you if I can't have a taste. From where I'm standing, your figs smell like your camel pissed on them."

"My camel is very well behaved. He never urinates."

But enough about early advertising. One thing the Net is good for is organizing markets. Especially if you're invisible and powerless, ignorant of how
things are supposed to work, ignorant of business-as-usual. Especially if you're
intent on end-running the empire.

Who has the stuff we like? Who makes the stuff we need? Interest, curiosity, craft, and voice combine to create powerful self-organizing marketplaces on
the Web: "Figs here, delicious figs!" Or it might be a faster chip, an elegant bit
of code, a new idea, a joke, a line of poetry, a job. Stuff, as the digital world
has taught us, isn't always stuff. And coordinating how it gets distributed is
more like a cocktail party than a strategy session. Stuff gets around the way
word gets around. Along the same routes. Around the same obstacles. Though
motivated by altogether different principles than those driving business, this
is not as chaotic as it may sound, nor as inefficient. It's happening right now,
every day. It works. "Follow the money" may still apply, but to find the money
in the first place, follow the conversation.

In this book, we have tried to paint a picture of radical changes that are
taking place today, aided and abetted by the Internet. But to people who've
already lived in the Net for a while, these changes aren't perceived as radical
at all. They're second nature. On the Web page we asked people to sign in
support of the Cluetrain Manifesto, one comment was repeated over and over:
"It's about time!"

We've talked about the ideas you've just been reading with hundreds and
thousands of people online who don't ask for additional explanation. Yeah, they
say simply, damn straight. These are people who "get it," as the saying goes.
They don't need explanations; they already know how it works.

"But...but..." you may sputter, "those are just disgruntled 'Net-heads'—I read
all about them in TIME or TV GUIDE or SPORTS ILLUSTRATED or somewhere.
Those unemployable fringe types who never amount to anything anyway."

Don't bet on it. Here's a small handful of the radical organizations in which
people who signed the manifesto work: Bank of America, Boeing, Cap Gemini,
Cisco, Comcast, Compaq, Computer Sciences Corporation, Dow Jones, EDS,

Ericsson, FedEx, Fleet Credit Card Services, Herman Miller, IBM, Intergraph, Kaiser Permanente, Kellogg, Kinko's, KPMG, Levi Strauss & Company, Lucent Technologies, Merck, Microsoft, Morgan Stanley Dean Witter, New York Life Insurance, Novell, Ogilvy Public Relations, Oracle, PeopleSoft, Pricewaterhouse-Coopers, Qualcomm, Saturn, Sears, Sema Group, Siemens, Sun Microsystems, US Interactive, the U.S. Internal Revenue Service, US West, USWeb/CKS, Wang, WR Hambrecht + Co., Ziff-Davis.

Stereotyping is a bitch, ain't it? Clichés are so comfortable and easy. Business is fat-cat moguls meeting in posh boardrooms atop steel-and-glass towers high above the jostling masses in the street. Stereotypes usually have some basis in reality, but they're lousy tools with which to frame critical judgments. More often than not, business happens in the streets, not above them. And so do revolutions.

But if you're looking for Molotov cocktails and tear gas, beleaguered cops and firebrand radicals, you're bound to miss what's really happening. Ruth Perkins of the Florida Department of Law Enforcement wrote to us, "Thank you for solidifying the thoughts and mission I've had for so long. I'm a wholehearted signer and practitioner of your manifesto."

Just because you're not seeing a revolution—or what Hollywood has told you a revolution ought to look like—doesn't mean there isn't one going down.

The Demonic Paradox

> *Although a system may cease to exist in the legal*
> *sense or as a structure of power, its values (or anti-values),*
> *its philosophy, its teachings remain in us. They rule our thinking,*
> *our conduct, our attitude to others. The situation is a demonic paradox:*
> *We have toppled the system but we still carry its genes.*

Ryszard Kapuscinski, Polish journalist, 1991

ALL TALK OF REVOLUTION NOTWITHSTANDING, THE struggle is already largely over. It's genuinely tough to find anyone who will stand up and defend the standard traditional conventional old-school way in which "everyone knows" business should be conducted. As far as we can determine, not only does everyone *not* know it, nobody seems to believe it for a second.

This is odd, we think. And critically important to us, personally and professionally. After all, if we're hanging our asses out with this whole CLUETRAIN tirade, there better be something there to carry on about. Right? Otherwise, wouldn't we look stupid?

So we rack our brains. We search our souls. We ask ourselves: Are we making this stuff up? Is it wishful thinking? Are we maybe just having acid flashbacks? Ever uncertain of our findings, but always wishing to be scientifically precise, we're all constantly performing little sanity checks: "Have I slipped the surly bonds of earth, or is it actually possible that nobody left alive today really believes this stuff anymore?"

We meet a lot of people in our day-to-day work. A lot of different kinds of people—as random a sample as you could ever hope for. Unbeknownst to them, they are being used as subjects—fodder if you will—for our ongoing market research. This involves looking for the perfect Suit, that is to say, the business person who fully embraces and embodies the corporate stereotype. So far, the closest we've come is some guy in a Dell TV ad: manly but understanding, firm yet gentle with his underlings. Always ready for a good laugh, but no joking around when it comes to delivering the goods. What he really does is hard to tell, though it seems to have a lot to do with his Inspiron-brand notebook computer. Man, he takes that baby everywhere!

But of course, he's a male model. So we're still looking. Most of the people we run across are rather disappointing in this respect:

"So how's the job going?"

"The job? How do you *think* the job's going? The job sucks."

"Oh."

Or maybe it's someone who just bought a new product online:

"Are you satisfied with your latest purchase?"

"What, are you yankin' my chain? Get away from me, you pervert."

"Yes, sir. Sorry to have disturbed you."

This is hard work. No lie. But we keep at it, relentlessly searching for the canonical business type or the ideal consumer. Neither seems to exist. Isn't that just too weird?

But here's something weirder still. If you take someone you've just been talking to in a normal, non-insane sort of way, and put him or her in a trade-show booth, nine times out of ten this person will immediately start talking like a Suit: "...and we are very proud of our preeminent position with respect to our competitors. Dunderhead & Gladhand just ranked our company second in the entire industry and..."

...and it makes you want to go out and shoot yourself, or at least take a long hot shower. Then he or she comes offstage and says, "So how did I do?"

You hem and haw. You want to be kind, but how to put it? "That was total bullshit! How could you spout that patent crap? I know you don't believe a word of it."

"Oh, that, of course not. But how did I do?"

Mr. Kapuscinski, our Polish journalist from the quote above, says that although we may have toppled the system, we still carry its genes. He says it's a demonic paradox. Jazzman Rahsaan Roland Kirk has another term for this same phenomenon. He calls it *volunteer slavery*.

So while business stereotypes are largely empty, or come from another day and have long since lost any real descriptive power, we find ourselves replicating the behaviors they caricature. Why? Well, because we're business people, of course! And that's how business people behave. Welcome to the hall of mirrors. Welcome, as Vonnegut put it, to the monkey house.

We don't believe what we're saying at work. We know no one else believes it either. But we keep saying it because because because because the needle's stuck. The record's broken. Because we just can't stop. Because who would we be if we didn't talk like that?

Maybe we'd be free. Or freer, at least.

In most cases, no one is forcing us to replicate these useless obsolete behaviors. We imagine we must, but we never investigate. We posit some organizational bogey man who'd punish us terribly if we were human. Give us a good hard whippin', you betcha.

What if there's nobody there, though? What if it's like Santa Claus, or flying saucers? Like Fox Mulder, we want to believe, we really do. Maybe it's like—uh-oh—God!

Not to be disrespectful, but there's a point here. Historically, capitalism depended heavily on the Calvinist notion that news of impending salvation was telegraphed by worldly success. Worker productivity positively skyrocketed under this inspired setup. It wasn't Santa who knew if you were naughty or nice, it was the Big Boss. So better knuckle down.

But, let's get real. A couple of centuries ago, a new invention arrived into the world. It was called "the job." The idea was simple, really. You went to some hellhole of a factory, worked sixteen hours until you were ready to collapse, and you kept on doing that every day until you died. Cool, huh? You can see where Calvinism must have come in handy. Some people wouldn't do that even for stock options.

Among the many casualties of this arrangement was the human spirit. And of its necessary functions, conversation was the first to go. People would talk with each other while doing craft or cottage work. But talk interfered with factory production. And of course, there was Management. Management knew everything. Workers knew nothing. So shut up and get back to yer lathes and looms, ye dirty sods!

Fast-forward a hundred years or so and along comes "knowledge work"—an even cooler invention that enabled us to have magazines like FAST COMPANY and meant we were allowed to know something all of a sudden. Excuse us, management said, but would you mind letting us in on whatever it is, as we're rather tapped out over here?

And the rest, as they say, is history. A history that brings us right up to today with its rip-snortin' high-speed Internet with broadband everything, hold the mayo. Whoopee! But that's not the point. The point is what this latest technological wonder brings back into the world: the human story. A story that stretches back into our earliest prehistory. A story that's been in remission for two hundred years of industrial "progress." When it breaks out again in the twenty first century, it's gonna make Ebola look like chicken pox. Catch it if you can.

And next time you wonder what you're allowed to say at work, online, downtown at the public library, just say whatever the hell you feel like saying. Anyone asks you, tell 'em it's OK. Tell 'em you read about it in a book.

Put that in your demonic paradox and smoke it.

More About Radishes

What do I know of man's destiny?
I could tell you more about radishes.

Samuel Beckett

SO WHADDYA THINK? WILL CLUETRAIN BE THE NEXT BIG Thing? Not if we can help it. Deep-six the bumper stickers. Forget the catchy slogans and the funny hats. Let's not write the bylaws and pretend we did. Let's not start another frickin' club. The only decent thing to do with Cluetrain is to bury the sucker now while there's still time, before it begins to smell of management philosophy. Invite the neighbors over, hold a wake. Throw a wild and drunken orgy of a party. Because only death is static. Life moves on.

How do you speak in a human voice? First, you get a life. And corporations just can't do that. Corporations are like Pinocchio. Or Frankenstein. Their noses grow longer at the oddest moments, or they start breaking things for no good reason. They want to be human, but gosh, they're not. They want the Formula for Life—but they want it so they can institutionalize it. The problem, of course, is that life is anti-formulaic, anti-institutional. The most fundamental quality of life is something the corporation can never capture, never possess. Life can't be shrink-wrapped, caged, dissected, analyzed, or owned. Life is free.

And so, finally, the question we've all been waiting for. In the newly humanized and highly vocal global marketplace the Internet has helped create, can corporations survive at all? Not if they're unable to speak for themselves. Not if they're literally dumbfounded by the changes taking place all around them.

But maybe—and it's a big maybe—companies can get out of their own way. Maybe they can become much looser associations of free individuals. Maybe they can cut "their" people enough slack to actually act and sound like people instead of 1950s science-fiction robots. Gort need more sales! Gort need make quota! You not buy now, Gort nuke your planet!

Easy there, Gort. Calm down boy. Here, chew on this kryptonite.

Everybody's laughing. No one gives a rat's ass. So here's another question. Perhaps you even thought of it yourself. How come this book ended up in the business section of your local bookstore instead of under Humor, Horror, or True Crime? Hey, don't look at us.

Fact is, we don't care about business—per se, per diem, au gratin. Given half a chance, we'd burn the whole constellation of obsolete business concepts to the waterline. Cost of sales and bottom lines and profit margins—if you're a company, that's your problem. But if you *think* of yourself as a company, you've got much bigger worries. We strongly suggest you repeat the following mantra as often as possible until you feel better: "I am not a company. I am a human being."

So, no, at the end of all this we don't have a cogent set of recommendations. We don't have a crystal ball we can use to help business plot its future course. We don't have any magic-bullet cure for Corporate Linguistic Deficit Disorder. Did that much come across? OK, just checking.

However, we do have a vision of what life could be like if we ever make it through the current transition. It's hard for some to imagine the Era of Total Cluelessness coming to a close. But try. Try hard. Because only imagination can finally bring the curtain down.

Imagine a world where everyone was constantly learning, a world where what you wondered was more interesting than what you knew, and curiosity counted for more than certain knowledge. Imagine a world where what you gave away was more valuable than what you held back, where joy was not a dirty word, where play was not forbidden after your eleventh birthday. Imagine a world in which the business of business was to imagine worlds people might actually want to live in someday. Imagine a world created by the people, for the people not perishing from the earth forever.

Yeah. Imagine that.

Journalism as a Conversation

DAN GILLMOR

I REMEMBER THE DAY I SAW THE CLUETRAIN MANIFESTO FOR THE first time, because the opening line, "Markets are conversations," gave form to a notion that had been growing in my subconscious for some time by then. And it was a small step to put that idea into words.

If markets are conversations, I realized, so is journalism.

I'd been involved in that conversation for some time by 1999, largely in my work at the SAN JOSE MERCURY NEWS, Silicon Valley's daily newspaper. But my understanding that the online world was all about conversation and collaboration long predated my 1994 arrival in California.

In the mid-1980s, having been online for a few years using early bulletin-board systems and commercial networks such as CompuServe, I had one of those epiphanies that comes to everyone who eventually recognizes the power of the network. I posed a small programming question to users of a CompuServe forum and got excellent—and different—answers from a variety of geographies.

When I took a job as an editor and then a columnist in San Jose, I discovered something quickly: My readers knew more than I did—and were happy, largely via email in those days, to share their knowledge (sometimes in the form of harsh but nonetheless valuable critiques of my work). Long before I had a blog, I was having an online conversation with my readers, and my work was immeasurably better as a result.

But CLUETRAIN's appearance, around the time the first blogging software arrived, catalyzed my thinking and changed the way I worked. Not only did it make me understand more clearly how journalism was changing, it also spurred me to push farther and faster into this new world than I'd contemplated earlier.

The idea of journalism as conversation inexorably leads to this understanding: The craft is changing for all three major constituencies—the journalists, the people and institutions journalists talk about, and most importantly the people I call the "former audience," who once were mere consumers but who now can (and in many ways need to) play a more active role.

Journalists are having to learn new tools and techniques, and include the audience in the process. The newsmakers can't hide as easily behind traditional PR or pronouncements. And as consumers become activists, they push the journalists to do a better job, and can even become journalists themselves if they choose.

The traditional media's long-standing opaqueness and arrogance are well recognized at this point by media customers, including audiences and advertisers. Those qualities mirror the ones that have kept so many businesses from fully understanding the nature of the CLUETRAIN changes.

Entertainment media and journalism both rely on scarcity-based business models, but have responded somewhat differently. The entertainment cartel of the past used every arrow in its considerable quiver, first to deny what was happening and then to stop its effects. Going online with big-selling music and movies was seen as inevitable. Understandably, the industry wished to preserve the relationship it had established with artists and the public alike as a middleman with the power to launch (or kill) artists' careers and set mass market tastes.

How did newspapers, magazines, and broadcasters respond to changing conditions? Most embraced the online world to some degree, but few recognized that online could provide more than just a new platform to distribute traditional content.

In fact, the word "distribute" is part of the problem. Media people still tend to think in terms of distribution. We create finished content and send it out, whether in trucks or over the airwaves—a mass, one-to-many approach that worked well in an era of information scarcity.

Today, it's not about distribution. Today, we create stuff and *make it available*. After we make it available, other people come and get it. Certainly there's an element of distribution—the act of getting our work noticed. But the fundamental nature of consumption has shifted from push to pull, or the modern hybrid that has emerged in the early days of conversational technology.

This shift fits perfectly into the CLUETRAIN universe, not just because the people we once called the audience are now part of the process, but also because consumers who become creators also become more savvy in the way they see and understand journalism. Some of the creators become collaborators, leading to a journalistic arena that is essentially unprecedented and, so far, underexplored.

My own blog began in the fall of 1999, half a year after the CLUETRAIN MANIFESTO was posted. I'd been watching the format develop, without entirely understanding how it worked. When Dave Winer, developer of one of the early blog applications, urged me to try one out, I did and was immediately hooked.

Blogging is the clearest example of media as conversation. A good blog has an identifiable voice. A human being is talking, not at us but with us.

Blogs subvert the old styles of prose favored by media and companies because of the voice requirement. Real people don't say to each other, "Three people were injured, one seriously, when two cars collided at the corner of Main and Elm streets." They say, "Did you hear about the crash over on Main Street? It was bloody."

Better corporate blogs subvert (thank goodness) the press release style. Read a standard press release out loud. It sounds as though it was written by the mate of a Turing Machine and a lawyer. It's an insult to prose and invites cynicism on the part of the reader.

When people belittle the small blogs—the ones written for family and friends—they are missing the point about why the genre has grown so rapidly. Surely the value, per reader, of the blog read by a few close friends is vastly higher than the value, again per reader, of the most widely read blog on the web. Random readers may enjoy what they find; friends care.

Blogs are part of a larger conversation about ideas and people. We in the blogging world respond to each other via links to the original piece. The hyperlink itself, a critically important part of the conversation, becomes part of a great autonomous machine that we have all created—the back-and-forth not just of a debate but our intentions when we share a link with our readers.

I was mostly alone as a journalist-blogger in the early days, though many people who weren't part of the traditional journalism world were doing work that had a journalistic feel and followed at least some journalistic principles. It took five years or more before newspaper blogs become even relatively commonplace; today it's a rare paper or broadcast outlet (or network) that doesn't have a variety of blogs written by staff members.

Why people took so long to recognize the value of journo-blogging eludes me. It was an absolutely ideal adjunct to my own work as a columnist, a way of trying out new ideas, alerting people to what I was looking into, and in general inviting my audience to share their own ideas. They had plenty.

Over time, I tried to take a somewhat open-source approach to my work. I didn't tell the world prematurely if I had a scoop (though in at least one case we were happy to break it on the web, rather than waiting for the newspaper to be published), but I did ask readers to help me figure out whether what I thought I knew was actually correct, or whether I'd missed something important.

This approach became a key element of my book on the citizen media phenomenon, WE THE MEDIA. In perhaps my most Cluetrainish move, I posted a long outline online, and then posted chapter drafts, all in the blog. The help I got from readers was extraordinary.

Traditional journalism's rapid decline as a business in the past few years has not been the result of bloggers taking away audiences, though some journalistic bloggers have earned serious readerships. The problem is more basic. It's about advertising, about money. Newspapers are trying to move from monopoly positions to a competitive market; broadcasters are moving from oligopoly to competition. Both are trying to do this in a world where people seeking information and entertainment have vastly more choices than before, and so do the people who have been paying the bulk of the bills—the advertisers.

The various revenue streams that have carried the freight for most newspapers in the past half century, notably classified advertising, have been under systematic attack from online competitors. They have won not by being cheaper (though most are), but by doing a vastly better job for both the advertiser and the potential customer. And the best of these enterprises do so in large part by creating communities, not just classified advertising tools. Look at craigslist.org, the famed site that Craig Newmark began as a mail list in San Francisco for his friends. It is substantially about community; the users of the site, not the employees, are the first line of defense against personal misbehavior and seller scams.

Contrast this with the way newspapers have sold advertising for decades. They wait for the phone to ring and then treat advertisers the way monopolists have always treated their customers, with contempt bred from complacency. Customers in a CLUETRAIN world don't put up with this. They demand a say in the transaction, and if they can't find a better solution elsewhere they're free to create their own way of selling.

There is little chance that traditional journalism organizations, with a few exceptions, will pull out of their economic tailspin anytime soon. The economic woes that become apparent in 2007–2008 created a nearly perfect storm for the media business.

Sadly, there's also insufficient evidence that traditional journalists will break free of their standard ways of doing journalism in time to make a serious difference. The press is a black box into which reporters push interviews and research, and from which emerge articles. Explaining how journalism is created? It happens, but mostly when a news organization screws up so profoundly that it feels compelled to come clean about what it did wrong; witness the Jayson Blair debacle, after which the NEW YORK TIMES published thousands of words describing how it occurred and how it planned to make sure such a thing wouldn't happen again.

Bloggers and others in the conversational media field have their own self-correcting system: readers. Not all bloggers correct their mistakes, but the smart ones watch what readers say in the comments and on other sites. The feedback system doesn't always work in a timely way, of course, and some bloggers ignore what they're told even when they're wrong.

A promising, if still minor, trend in traditional journalism (online and offline) has been to ask the audience for help doing the reporting. This takes many forms. The BAKERSFIELD CALIFORNIAN newspaper has asked its readers to fill in a map of the community's street and highway potholes. A newspaper in Florida, confronting policy changes in regard to sewer rates, asked readers to help figure out what was going on, and produced investigative journalism that led to changes in those policies. Blogger Joshua Micah Marshall, founder and editor of TALKING POINTS MEMO, has won major journalism awards that formerly were the province of print publications. He has repeatedly asked his readers to help gather information for stories.

One of the least conversational parts of the newspaper business has been the way publications handle their archives: hide them away behind what Doc Searls aptly calls "pay walls," asking for payments when people want to go back into a community's history. This is insane, and in the end it's an almost completely counterproductive policy. Doc and I, among others, have urged newspapers to put their archives into the public sphere. So far, most have not. One reason is contractual obligations to online libraries, but another is pure myopia.

Local newspapers, for practical purposes, archive their communities' histories. Certainly there are gaps, but on almost every issue of importance, and when it comes to tracking what prominent local people have done in their lives over recent decades, the newspaper's archives are where you'll find the most comprehensive accounts. On just about any local topic you can name, the newspaper has done stories that would be high in any Google search on that topic. Considering that traffic to those articles would soar if they were visible, it seems likely that newspapers would reap financial gain by making archives more open.

One prominent newspaper that has put its archives out in this way is the nation's top publication: the NEW YORK TIMES. It is safe to say that in a decade there will be a TIMES story link in the first page of any search you make on just about any topic or person of national or international importance, assuming the TIMES covered the person or issue in its standard depth. The paper's parent company is undergoing plenty of financial stress for other reasons, but its archives are almost certain to be a long-term generator of cash, as well as making the TIMES the go-to place for vital background on countless topics. That's a weblike way of doing business.

Journalism as conversation is the linchpin to the rise of citizen journalism, which takes many more forms than can be described in this short essay, for example, the photo posted by a nonprofessional from a major event, such as the iPhone camera shot in January 2009 by a man on a New York City ferry boat as it approached a US Airways jet that had ditched in the Hudson River; "placeblogs" that understand and cover a community too small for traditional media to care about (except when something bad happens); neighborhood mail lists where people tell each other about things that are genuine news; and so much more.

What we create ourselves does not exist in a vacuum. It exists as part of an emerging journalistic ecosystem that I hope will be more diverse and robust. Traditional journalism should, and must, survive. But it needs the help and the competition from the new players.

For professionals and so-called amateurs alike, the tools continue to evolve. A decade ago, not only was blogging new and the notion of journalism as conversation almost inconceivable, but the words "podcasting" and "Twitter" hadn't entered the language. Few guessed that millions of people would soon upload videos to the web and then comment on each others' offerings. And in the Web 2.0 era, we combine data and software services from a variety of places with our own ideas and information, in a much more interactive way than we contemplated in 1999.

Audio and video remain the least conversational of tools, largely because they're linear by nature. Still, entrepreneurs and big companies alike in the field are finding ways to increase the interactive possibilities.

Photos have become extraordinarily conversational, however, in part because of services like Flickr, which are all about sharing. The sharing economy—so well explained in books such as Yochai Benkler's THE WEALTH OF NETWORKS and Clay Shirky's HERE COMES EVERYBODY—is entirely about conversations. In countless cases, citizen journalists upload their work to the world long before the traditional media are anywhere near the scene.

I find myself increasingly intrigued these days by Twitter, which I've called "a current (and sometimes early) warning and gossip system, with occasional insight and frequent amusement." Twitter's microblogging nature, especially its limit of 140 characters per posting, requires brevity and encourages wit—ideal qualities of a good conversation.

The conversation around 140 characters extends, of course, to a medium we'd barely begun to use this way a decade ago, at least in the United States: the mobile phone. People in Japan and Europe, especially the Nordic countries, were already texting each other back then. Text has become the primary mode of digital communication in Africa, and we in the developed world are learning from how people use text in places where, so far, they have few other alternatives.

In the end, however, the ultimate journalistic tool is the human brain: our ability to learn and absorb and adapt. In a conversational mode, we do all of those more effectively.

And what is the first rule of a conversation? To listen.

Traditional journalists were poor listeners in the era of monopoly or oligopoly media. But at long last and in no small part due to the ideas that sparked CLUETRAIN, they are relearning this vital lesson. Let's hope we're all doing the same.

How Lego Caught
the Cluetrain

JAKE MCKEE

IN THE CLOSING PAGES OF THE CLUETRAIN MANIFESTO, the authors make a point so profound that it can only be described as odd: "The only decent thing to do with CLUETRAIN is to bury the sucker now while there's still time, before it begins to smell of management philosophy." Considering that turning a core idea into a "management philosophy" is what most business books attempt to do, you can see why I'd use a word like odd. But perhaps this approach wasn't odd at all. Perhaps the reason their book took off is precisely because they chose a more complex and thought-provoking path.

In 1999, I read the ninety-five theses that the authors tacked to the Internet, as they like to say. In 2000, I bought my first copy of the print version of THE CLUETRAIN MANIFESTO, reading it cover to cover twice in the first week. The points that I had tried to make to my web development clients suddenly had a voice, a vocabulary, a point of reference.

But like the CLUETRAIN authors, I became increasingly nervous as I heard more and more people parroting a line like "markets are conversations" without fully understanding what it means or how to apply it. It was too easy for marketers to cloak their old business in the new talk with doing anything new or different.

But just as we began to see some of these fears begin to play out, I was given a chance to apply the CLUETRAIN thinking to a company woefully and sadly behind the times—a company everyone knows, that many love, that I had obsessed over for twenty years. I'm talking, of course, about the Lego Company, the toy maker responsible for those iconic plastic building bricks. Founded in 1932 as a wooden toy manufacturer by a Danish carpenter, the company began the path to becoming a global cultural icon in the 1950s with the addition of the plastic Lego brick to their product line. Since then, wooden Lego toys have become a thing of historical record, while today's kids spend upward of 5 billion hours a year playing with Lego bricks.

To understand the CLUETRAIN story at Lego, you have to understand my personal story as an AFOL (adult fan of Lego). And to understand my story, we first have to travel back in time to the early 1980s.

Growing up, my friends aspired to be firemen, policemen, and astronauts. Very cool professions, to be sure. But they were only marginally interesting to me. When adults asked young Jake what he wanted to be, Jake always responded, without hesitation, "A Lego designer!"

For so many of us growing up in those days, Lego wasn't just a toy but a tool of our imagination, a tangible expression of the visions we had in our head; it wasn't just a big pile of bricks but a big pile of bricks waiting to be formed into something amazing. The "brand promise" translated from "you can build any-thing" to "you can be or do anything." After all, if you could make a TIE Fighter out of the white, rectangular shapes, who could tell you that you couldn't do anything else in life? Whether I realized it at the time or not, this is when I first learned that I didn't buy the "brand." I bought what the brand promise helped me imagine I could do. Ironic, considering that ten years later I find myself leading my own company focused on helping brands activate their promises through conversation marketing. (I'm still trying to decide if the CLUETRAIN authors would be proud of the new company or bothered that I've turned their ideas into management philosophy.)

By the mid-1990s, the Internet had started to become a powerhouse in pulling together niche groups from all over the country and all over the world.

Interests long forgotten could be rediscovered as easily as a web search or an eBay purchase. Adults who had grown up on Lego, many of whom had never really stopped building and playing, started to find each other, sharing pictures of their Lego creations, discussing building techniques, exchanging building elements, and sometimes even meeting offline. Lego's initial brand promise had returned: If you were interested enough in the hobby, a world of opportunity waited for you.

I was one of those who began meeting, sharing, and interacting with other Lego fans from all over the world. If you can imagine the stigma of adults playing with kids' toys, you can also imagine the comfort that came with finding your tribe. And as more and more adult Lego fans connected, the more they wondered openly (and loudly) why the company they loved so actively resisted engaging with them.

By 2000, when THE CLUETRAIN MANIFESTO first appeared in print, the brand promise that so many AFOLs had grown up believing was now something much more grand. People were realizing that they wanted, needed perhaps, to interact with the companies who were supposed to help deliver on their own brand promises.

But Lego, the company, was far, far removed this spirit. In fact, it was so far from the core business that when the Lego staff visited the biggest three customers (Walmart, Target, Toys R Us) in 1999 to talk about the upcoming product line, all three customers independently questioned the lack of understanding Lego had about the Lego buyers themselves.

Welcome to Fort Business.

I joined the Lego Company in late 2000 as part of the LEGO Direct team, a new division responsible for managing all direct interaction, sales, and marketing efforts that reached the customers, well, directly—an initiative that served as the company's response to its distance from the market. In a company primarily built around distributing product to large, big box retailers, direct interaction was surprisingly far from the core business. Yet it spoke immediately to the spirit of the AFOLs.

The Lego executive team had been aware enough of the shift in consumer expectations to launch a group like Lego Direct, but the company overall had yet to fully understand, much less embrace, the ideas in CLUETRAIN. The company had for years developed a culture that largely rejected, or at least ignored, outside input. And the company almost exclusively focused around kids, not these "weird adult fans" (you know, the ones with passion and, um, money).

As an AFOL myself, and as an AFOL who had seen the impact a small group of adult fans can have on large audiences of kids and families, I wanted to see this change. Fortunately, so did my boss and head of Lego Direct, Brad Justus. Not only was Brad a supporter of the Cluetrain, he had specifically designed the business unit around the principles outlined in the book. Brad once told me:

CLUETRAIN was a concise and perfect manifestation in words of something I had been considering for some time: the notion that the ideal enterprise (company, or other undertaking) embraced customers and employees equally; that the old paradigm that the company/customer relationship was one of parent to child. 'You'll eat what I give you' was hopelessly outdated, and technology only accelerated the move away from that paradigm. In fact it underlay an entirely new form—one of enterprise as vast social network, a "real" network which was not necessarily the now more rampant virtual ones, where all participants play interdependent roles and all contribute equally to the success of the venture (the growth of the company, the creation of the ideal product, the spread of the brand promise).

Looking back on it, I had taken to heart lessons from the manifesto that helped change the way this decades-old company approached its adult fans, as well as its entire customer base. It's easy today, with the company fully embracing its community of adults, kids, parents, and families, to look at the ninety-five theses and map many of them directly to the work our team did inside the company over the years.

52 **Paranoia kills conversation. That's its point. But lack of open conversation kills companies.**

Events and decisions inside a company can seem innocuous at the time but can lead to hugely detrimental unintended consequences. In the 1980s Lego started receiving more and more letters from kids, many of which offered up their own product ideas. This was also the time when lawsuits were being filed at an increasing, somewhat alarming rate.

Whether out of fear of legal threats or actual legal threats (I was never clear on this), the company became legitimately concerned about protecting itself from kids who sent product ideas into the company and then saw their idea on the shelves a short time later. Imagine a kid who wrote in with a great idea like "I think you should make an underwater theme," only to see it on shelves months later. (Keep in mind that it took upwards of eighteen months to develop themes at this time.)

The Lego legal department issued a directive that the company would no longer accept "unsolicited product ideas." A logical, rational, yet eventually damaging directive.

The key to this directive was the word "unsolicited," meaning random submissions required active rejection. If you mail in an idea, the directive dictates, Lego should send it back to you with a "thanks but no thanks" note. It was that simple.

Over time, the specifics of this directive became a bit fuzzy. People lost track of the original intent. By the time the Internet and email came around and the quantity of input increased exponentially, the culture at Lego was already steeped in paranoia. Not only did the company not want to accept incoming communication, but it didn't want to let anything out either.

Lego Direct had to redefine the concept and purpose of secrecy.

It's easy, perhaps comforting, to simply lock down everything, not allowing any content, ideas, or discussion to leave the virtual four walls of the company. You can't get in trouble for what you *don't* say, so why say anything at all? What our team found, though, was that when faced with the question, "Okay, so tell me *why* that's secret," colleagues couldn't answer it (most of the time, anyway). Secrecy was comfortable, and we had to pull them out of their comfort zones. That required executive (or at least managerial) support, and we found that it was surprisingly easy to get. Assuming, of course, that we had the tenacity to actually ask—something that didn't come easy for far too many of our colleagues. It's hard to argue for comforting secrecy against the potential business gain.

We also worked very hard to eliminate the use of nondisclosure agreements (NDA) wherever we could. In theory, these agreements ensure that you are able to share private information with outside partners without fearing that that the information will be distributed inappropriately. Knowing that most information isn't really "secret," and knowing that NDAs inherently indicate a fundamental lack of trust, we worked hard to protect ourselves in other ways. We found that working with the right people, empowering them to represent the community, and giving them data that made it clear that we trusted them was more useful than any piece of paper.

In fact, the perception of trust helped us to create vastly more safe interactions with the outside world, which in turn helped pull the company out of the embedded culture of paranoia it had been rooted in for so long.

60 Markets *want* to talk to companies.

In December 1999, Brad Justus, Lego Direct's first executive leader, posted an introduction to the adult hobbyist forum, LUGNET (http://news.lugnet.com/general/?n=11596),

> To LEGO enthusiasts everywhere:
>
> We hope that we are the bearer of some holiday cheer for you all. For those of you who have felt that your love for LEGO was unrequited, this is a new day. With this missive, the LEGO Company asks to open a dialogue with you, our consumers. Whether you are an AFOL, or a parent purchasing a first DUPLO set, or a KABOB (Kid with a Bunch of Bricks–we just made that up), here are some words that should gladden your hearts: We are listening.

This message received more than two hundred replies, almost all positive and welcoming. More than two hundred replies from a community of, at that time, of a few hundred people.

(It was during this same time that I began my quest to get a job with Lego

Direct, and Brad's message motivated me to believe that the company was porous enough for me to actually get inside.)

Most LUGNET users were thrilled. After years of ignoring the AFOLs, Lego had finally wised up and was now engaging the community directly. This sparked incredible new enthusiasm for the company. It also began to legitimize the hobby, reinforcing the idea that maybe it wasn't so weird to play with "kids toys" after all, and it planted a flag in the ground saying, "We're listening."

After joining the company several months later, I was tasked with meeting as many fan groups as possible. Surprisingly, these first meetings consisted mainly of me listening while the fans complained about the lack of communication from the company in the past decade.

But after these initial meetings, the relationship began to find a firm foundation, based primarily on the channels that had been opened, first by Brad's note and then by the real action that followed. In the days before blogs and common access into the company, the fact that our team monitored and interacted with fans through forums, emails, IM, and personal phone calls was quite a revolution.

But don't get me wrong; the AFOL community wasn't waiting on the Lego Company. Far from it.

85 **When we have questions we turn to each other for answers.**

It may have taken until 2000 for Lego to participate in the fan community, but by that time the AFOLs had already created a vibrant community ecosystem. By the late 1990s, they had created a photo sharing site (brickshelf.com) with tens of thousands of Lego-themed photos posted, an incredibly robust discussion forum that (lugnet.com), an inventory system that enables users to find out every part in a particular Lego set (peeron.com), and a virtual shopping mall where fans could open a store front and sell individual elements (bricklink.com).

The mistake many companies make when they first engage a community is to rush in and try to replace unofficial efforts with official efforts. Even if such a move is well intentioned, it's as if the company is saying, "Your efforts are sub par. Let us professionals step in and show you how it's done." Not a very good way to start off the relationship.

It also fails to appreciate the reality of the enormous amount of work it would take to replicate fan-based efforts. If Lego had tried to create a replacement official photo-sharing site to replace Brickshelf, we would have faced a monumental task. In addition to building an entirely new website at a significant cost, we would have taken months to even begin to scratch the surface of the quantity of images already posted to Brickshelf.

Our task with these community projects was to embrace and help support them where possible. If these community sites needed financial support, we often gave it. If they needed our participation to gain legitimacy, we showed up. If they needed goodies for promotions that generated user engagement, we dropped things in the mail.

The more we respected their efforts, the more they asked us to participate. And the more we participated in their community, the more interested they become in helping us with ours. We were learning about the fair exchange of value—a fundamental in community engagement. In time, this even helped me develop the mantra I use to this day: Everybody goes home happy.

72 We like this marketplace much better. In fact, we are creating it.

As I write this, Bricklink.com, the virtual shopping mall for fans to sell Lego parts, sets, and collectibles among themselves, has more than 84 million items for sale from more than three thousand storefronts. The owner of the site takes a 3 percent fee on each sale and is making enough money to live off of site revenues. Many of the storefront owners are making a tidy living from the sale of individual Lego elements. And the overall quality of model design has increased exponentially as parts have become easier to acquire.

The brand promise I grew up with—I can build anything I can imagine—has been elevated to a place that no one could have imagined in years past. And the evidence of this brand promise comes from the users themselves instead of marketing materials produced by the Lego Company. Fans have embraced the promise and made it even more real than it was when I was a kid. They've embodied the Lego brand promise, something no company could make happen on its own.

But this new marketplace wasn't taking away opportunity from the Lego Company, nor was its influence limited to adult fans.

80 **Don't worry, you can still make money. That is, as long as it's not the only thing on your mind.**

In a few short years, Lego had gone from almost never engaging adult fans to having them participate in the development of multiple product lines and concepts. The rest of the company thought the Lego Direct team was crazy when we announced the launch of the $300, 3,000-piece Lego Star Destroyer model, targeted to builders sixteen years old and up. This was the largest, most expensive set in the company's history, and the most common reaction from our colleagues was "you're insane."

We had a hard time keeping them in stock.

By the time the Star Destroyer model was released, we had a wealth of knowledge collected primarily from two sources. The first source was our initial and ongoing experimentation with product releases. In the earliest days of Lego Direct, we had been releasing fan-friendly set designs. They started out much smaller and less expensively, but with each new set we released, we learned about what kinds of things excited fans of all ages. Second, we were consuming feedback as fast as we could. We were reading countless forum posts, having conversations with both individuals and user groups, and were asking product questions to any fan who would stand still long enough. By the time we produced the Star Destroyer, we had a firm understanding of how to excite and deliver.

Lego's entire product line has grown for the better, with sets for all ages going back to the basics of what has excited people of all ages for decades: fantastic models, creative ideas, and a brand promise that helps inspire kids to let their imaginations loose.

Today, consumers can buy individual Lego elements in any quantity they want. They can design their own models with the free Lego Designer software, upload the models to LEGO.com, and have the pieces for those models shipped to their house in a custom box. And in 2009, ten years after the CLUETRAIN was first tacked to the Internet, children, parents, and fans will be able to experience Lego in a way never before realized: a massive multiplayer online game called Lego Universe.

In 2000, the authors of CLUETRAIN published those immortal words, "Markets are conversations." The Lego Company is an organization that took those words to heart. While I have moved on, I can still see the spirit of engagement, of conversation living on in the product and in the enthusiasm that spills out from fans and parents and kids alike.

One day soon I hope to be sitting on the floor in front of a big pile of bricks with my daughter. Maybe she'll tell me that she has a great idea for Lego and I'll be able to tell her that she can tell them about it, knowing that they'll be listening.

Cluetrain in the Cubicle

J. P. RANGASWAMI

THE CLUETRAIN MANIFESTO. COULD IT BE TEN YEARS ALREADY?
Everything seems faster nowadays. Probably something they put in the water.

When I was asked if I would contribute something for CLUETRAIN's tenth
anniversary edition, the first thing I did was to try and remember where it all
started. And, as with so many things in those days, it began with RageBoy. I was
an avid reader of Entropy Gradient Reversals (as in the email newsletter of the
time), and I think that's where he mentioned what was happening with the
MANIFESTO. The website turned into a book, one thing led to another, and
before I could say "Rick Levine" I'd met the other three authors in different
places at different times. (In fact I still have a triple-signed copy of the book waiting for Rick's John Hancock. Rick, you have been warned.)

Where was I? Oh, yes, Bangalore. Yes, that's where I first met any of the
CLUETRAIN Four; we managed to convince RageBoy to come over and speak at
one of our development offsites soon after the book was published. Suffice it to
say it was an unusual experience, hearing Chris Locke speak, trying to get people
at work to buy in to what CLUETRAIN stands for.

Now, ten years on, what impact has CLUETRAIN had on people at work? Have things really changed? If so, how? If not, why not?

I think we need to bear in mind that CLUETRAIN was written in heady times of rapid growth and rapid change. Today the world looks a lot different, as we make our way through the detritus of a major global recession. On second thought, we don't have to bear that in mind—it is of no consequence except as a historical fact. Because, if anything, CLUETRAIN is more true today than it has ever been.

Why? To me CLUETRAIN is about many things, ninety-five things in fact. But more than anything else, it is about five things:

- It is a philosophy, a set of values, a mindset in which people rediscover their humanity.

- It is about the way people engage with their customers.

- It is about the way people engage with their colleagues.

- It is about the way people engage with people.

- It is about the changes that take place as people begin to engage with people again.

As markets became conversations. Again.

Peter Drucker once said, "Most of what we call management consists of making it difficult for people to get their work done." And I think that's the heart of the problem. Ronald Coase's 1937 Theory of the Firm was all about reducing transaction costs, taking out friction. What Drucker saw was the very opposite. Organizations were qwertying themselves, slowing themselves up by design. And calling it management.

CLUETRAIN pointed out why this was happening. We were building walls between us and our customers. Building walls even between us and our colleagues. Then we made it worse by using the most stilted and artificial modes of communication possible. And in the process, we were dehumanizing ourselves, our customers, our processes, our communications, everything.

To my way of thinking, it was all because of two popular ideas: the economics of scarcity and the assembly line. The economics of scarcity is, to all intents and purposes, basic economics to most people. Value is based on scarcity, and so scarcity itself is valued. If things are not scarce then they should be made scarce, artificially if required. Which leads to quotas and barriers and hoarding and butter mountains and you can imagine what else. The assembly line is all about division of labor, specialization, and reduction of standard deviation. When you put the two together, what happens is that you no longer remember that (to quote another Druckerism) "people make shoes, not money."

It is said that every economic era is characterized by a set of abundances and a set of scarcities. Firms that can take advantage of the abundances as well as the scarcities are bound to succeed. The trouble is, we've forgotten how to deal with abundance. And our natural response is to create artificial scarcities.

Hierarchies are about boundaries, about scarcities. Networks are about abundances, about relationships. And everything is about people, who are naturally more inclined toward network and relationship rather than hierarchy and boundary. CLUETRAIN was trying to point that out hyperlinks subvert hierarchies: "We are not seats or eyeballs or end users or consumers. We are human beings—and our reach exceeds your grasp. Deal with it."

Let's see how we've fared over the past decade. Let's start with how we deal with the customer. Finally, finally, it looks like we've stopped referring to customers as numbers, as "customer relationship management" systems have come of age. In general, we now refer to people by their names rather than their account numbers. This is a good thing.

But. As we've gotten rid of one evil we appear to have created another. We now have systems that "target" the customer, that "acquire" the customer. Systems that call themselves "customer relationship management systems," although no customer has access to them. Maybe we should call them by their rightful name—customer exploitation management systems—and be done with it.

There is a better answer, vendor relationship management, the corollary to customer relationship management. And it's coming. Not surprisingly, it's coming out of the CLUETRAIN stable.

For far too long, customers have been treated like slaves in this context. Targeted and acquired. Their freedom has been made artificially scarce, as firms seek to raise switching costs and migration paths through fair means as well as foul.

But the news is not all bad. Data portability is becoming something that customers understand the need for, and as a result firms are having to respond; being treated as human beings, names rather than numbers. With customer data beginning to be recognized as belonging to the customer (how did that happen?). With customers having more and more freedom to switch from one source to another as services commoditize, as healthy competition emerges, as the markets mature. Ten years on, the customer is not doing that badly.

What about the walls within the firms? Are hierarchies disappearing? I think so, driven by three almost independent factors:

- Generation M's entry into the workplace, the advent of mobile, multimedia, multitasking youth

- The inversion of the technology adoption curve; nowadays these kids try out new technologies far earlier than their predecessors did

- The growth of social media, software, and networks

You see, we could do things like insist on people using company PCs or phones. But that's a bit like telling people to restrict themselves to using company pens. It won't wash anymore. People have finally figured out that the letter P in PC stands for personal. When the youth of today stroll into work, their laptops and phones are theirs—reservoirs of their digital assets, their communications tools, their entertainment devices . Oh yes, and somewhere along the line they're also their PCs and phones.

We've tried telling them that our desktops and environments are "lockdown," that they must conform to the way we work. Or go somewhere else. And guess what? They're going somewhere else. As long as there's a war for talent, we need to understand how to tackle this. And artificial scarcity is not the way.

Much of what we call privacy and confidentiality and security is actually artificial scarcity, doomed to fail. There is only one way to keep a secret: Tell no one. As against this, the Internet is a great big copy machine, as Kevin Kelly reminds us. What happens when you connect something to the Internet? It becomes easier to make copies of things that are connected to the network in the first place. So people figured that out and tried to make it harder and harder to make the copies. This attempt at artificial scarcity was met by equal and opposite forces of artificial abundance and therefore failed.

That's what social media is doing, breaking up the artificial scarcities that have corrupted organizations. The concept of embargo, a critical component of old journalism, is a classic example of an artificial scarcity. Soon it won't be possible to have one.

Of course, the immune system of the enterprise fights back. Tries all the standard lies. *It's too expensive. It's too insecure. It's a waste of time.* But soon the lies are shown for what they are: lies.

Social software tends to have very low implementation costs as software per se. The cost, and the value, is in the way people use the software. What they contribute, what they share. When we introduced social software at one of the firms I worked for, it wasn't long before I got a classic email: *You have been encouraging your department to blog. I have been analyzing the blog posts of one of your subordinates. I notice that of the last fifty posts he has made, 47 have nothing whatsoever to do with work, 2 are possibly related and only one is clearly and undeniably to do with work.*

I resisted the temptation to write back; instead, I went up to the person and said, "I'm delighted that you had the time to analyze my colleague's posts. Could you please compare and contrast the conversations you overhear at the water coolers and at the washrooms on a similar basis? I am sure the output will be as useful."

Which reminds me. One of the ways we used to become subhuman at work is by writing memos. Instead of talking to each other, we would write notes. Even if the person we were writing to was a few desks away. If anything, email exacerbated this. Social software is bringing the balance back.

Not everyone understands this. There are many places where people think that blog posts should be treated as press releases, with workflow for approval prior to publication. Which just goes to show the extent of the education needed.

All in all, the advent of the new generation, coupled with the tools and technologies at its disposal, means that companies are less likely to be able to operate on a command and control basis.

Evidence of true empowerment of the individual within the firm is still limited. By this time, ten years after CLUETRAIN, I would have expected to see many of the day-to-day decisions taken at work to be devolved. This isn't happening.

There are a number of reasons, in particular information asymmetry, another example of artificial scarcity. People seek to obtain power by holding on to the information required to make decisions. Tools to support emergent behavior are still somewhat immature: We don't see prediction markets used widely within enterprises; even simple wiki cultures are slow to embed themselves.

Why is this? It's because people are incentivized not to share information. We speak a lot about collaboration tools and techniques, we speak a lot about knowledge management, we speak even more about collaboration. But, in the main, this is not what happens within the enterprise. When I worked in an investment bank, for example, it was nigh on impossible to get investment bankers to part with their little black book of contact names and addresses. It was their identity, the basis of their worth to the firm. And as long as it's what they believe, nothing will change.

All this goes beyond the incentivization problem. The very nature of the firm is changing. Firms used to be about siloed businesses and products and services, set in hierarchies, with locked in customers. Now, as Venkat, a friend who's a professor at Boston University is wont to point out, firms are collections of open relationships and capabilities. Firms used to be about access to capital, about global reach, about benefits. Now most people have better credit ratings than their employers. And benefits have become corporate millstones.

I think there are two other forces of significance that have been acting on the decade that's past, forces that have been paving the way for CLUETRAIN in the cubicle. The first is the customer's need for simplicity and convenience, as people begin to value their time more and more. This leads to a demand for self-service, which in turn puts the customer in control. Self-service paradigms are absolutely critical to CLUETRAIN, something I've understood more over the years in conversation with the authors. Do-it-yourself is a good thing for the customer. There is no artificial scarcity in DIY.

The second is the growth of open multisided platforms in business, based on deeper understanding of what Doc Searls terms the Because Effect. When something is scarce, you make money with it; the scarcity has a market value. When something is abundant, you make money because of it, not with it. Google and Amazon both make money because of Linux, not with Linux.

In summary: The walls are coming down. As things commoditize, and as the Because Effect rattles through industry, customers are finding their own voice again. Since all this is happening when members of a new generation are entering work, and since they have tools that are more collaborative than those of their predecessors, the scene is set fair for CLUETRAIN to enter the cubicle.

And particularly since all this is happening when we're in the trough of a deep recession worldwide, it becomes even more interesting. CLUETRAIN is about human beings connecting with each other in human ways. And it's what we need for the problems we face, not just at work but around us as well.

So I'm optimistic. CLUETRAIN will enter the cubicle. And we will learn to be Connected, not Channeled.

ACKNOWLEDGMENTS

THIS BOOK IS THE RESULT OF LITERALLY THOUSANDS OF conversations over the course of many years with friends, colleagues, online acquaintances, readers of our various 'zines, and the signatories of THE CLUETRAIN MANIFESTO. While we are grateful to all, it would be impossible to thank everyone individually. However, we owe a particular debt of gratitude to the following people who made especially generous direct contributions to the project:

Tom Petzinger, Jr. of THE WALL STREET JOURNAL, an early CLUETRAIN enthusiast and the author of the best business book of the past five years.

Dan Gillmor, who despite strong reservations about the MANIFESTO's style deemed it significant enough to cover it in the SAN JOSE MERCURY NEWS.

Bob Gahl, Dan Pritchett, and the team at The Sphere Information Services (www.thesphere.net) for hosting cluetrain.com.

Jacqueline Murphy, our unflappable editor at Perseus.

Dinah "Metagrrl" Sanders, who pitched in to help create and maintain the Cluetrain Web site.

Susan, David, Sarah, and Naomi Levine for letting the authors invade their aerie for our face-to-face meetings.

And we give a very warm acknowledgment to David Miller of The Garamond Agency, the "fifth engineer" who understood instantly what we were on about, and pulled us through the entire process with insight, friendship, and warmth. Thanks, David!

Rick Levine

QUOTING JP, ONE OF OUR ILLUSTRIOUS CONTRIBUTORS TO THIS edition, in April 2008:

> Rick Levine a real person? Next you'll be telling me that Christopher Locke exists. I had dinner with RageBoy only a fortnight ago, and he confessed to me that Chris was a figment of his imagination. As was Rick. That it was just his way to cope with the royalty-sharing agreements with the two Davids.

I've always been the unofficial black sheep of the CLUETRAIN quartet, more interested in making businesses successful than in finding time to write about making businesses successful. I don't get out as much as David, Chris, and Doc. Despite being dipped in too much technology since I was fairly young, I still like spoken, face-to-face conversations over mediated ones. Since CLUETRAIN was first published, I've continued doing what I enjoy most: working in and creating great, small teams focused on making stuff.

In early 2006, I was between start-ups. I'd been dabbling with chocolate and candy for years, mostly because I like working with my hands, enjoy cooking, and it's fun to do in the kitchen with my kids. My wife, Sue, challenged me to sell some chocolate at a fundraiser for her orchestra. (I was spurred on, in part, by my tendency to take over all her kitchen and pantry space with chocolate junk...) We sold fifty boxes of really good chocolate for $40 each in under an hour. In the next two months, Seth Ellis Chocolatier was born. Among my fascinations with the business is that it runs on word of mouth, on conversations. With a little bit of tasting thrown in.

One of the nicest side effects of my current gig is my kids now have a much easier time describing what their dad does for a living. They went from "he works with software and internet stuff and video and I really can't describe it" to "he makes chocolate." I like that.

Christopher Locke

AFTER CLUETRAIN APPEARED IN JANUARY 2000, PERSEUS PUBLISHED two other books by me: THE BOMBAST TRANSCRIPTS: RANTS AND SCREEDS OF RAGEBOY, and GONZO MARKETING: WINNING THROUGH WORST PRACTICES. It was with the stellar sales performance of these two volumes that I officially entered the American underclass. In 2005 (best as I can recall), I became Chief Blogging Officer for HIGHBEAM RESEARCH and also started blogging MYSTIC BOURGEOISIE. I still occasionally post the odd bit of flotsam to ENTROPY GRADIENT REVERSALS, which I originally started as an email newsletter back in 1995, when (for my sins) I was working at IBM.

Disclaimer: Nothing to disclaim at this time.

Doc Searls

DURING THE SUMMER OF '99, WHILE WE WERE BUSY WRITING THE first edition of this book, Dave Winer urged me to start writing a blog. Dave had created some of the earliest (and still the best) methods for writing blogs, but I was too busy to listen. Then, a couple months after we finished the book, I relented, and started writing something I called a "Cluelog." My fantasy was to have all four Cluetrain authors write the thing. But it turned out that blogging—at least in those early days—was more of a solo than a group activity. Chris and David urged me to make the blog my own, which I did. Both of them went on to write blogs as well, in addition to a pair of books apiece. Great ones, too. Go read them.

Blogging is a useful way to vet and scaffold ideas about subjects, and I've done a lot of that over the last decade. I've also been Senior Editor of LINUX JOURNAL for the same period—work that began when I started covering the Web for that magazine and others in 1996. I've been recognized and rewarded well for the work. In THE WORLD IS FLAT, Tom Friedman called me "one of the most respected technology writers in America." Google and O'Reilly co-gave me their Open Source Award for Best Communicator in 2005. (It was a cash prize too, which was nice. I can't be bought, but I can be paid.)

In 2006 I became a fellow at Harvard's Berkman Center for Internet and Society (where David Weinberger is a colleague) and at the University of California, Santa Barbara's Center for Information Technology & Society (CITS). My work at Berkman is focused on ProjectVRM, which I explain in my new CLUETRAIN chapter. My work at CITS is focused on the Internet as a form of public infrastructure. Both are filled with brilliant, wonderful people. As places to hang out while changing the world (or at least trying), one can't do better.

Besides all that, I do a lot of public speaking and run a business consultancy. It's a good life.

And yes, I am working on a book. It may not be the easiest thing for me, but it's important and I will get it done soon.

David Weinberger

THANKS TO THE SUCCESS OF CLUETRAIN, I WAS ABLE TO ALTER THE mix in my revenue carburetor (do cars even have carburetors any more?), spending more time writing, although I continue to do a lot of speaking and some consulting.

In 2002, SMALL PIECES LOOSELY JOINED: A UNIFIED THEORY OF THE WEB came out. It tries to explain why the weird world of the Web feels so familiar to us. [SPOILER ALERT] It's because the Web provides a better reflection of who we are than do other media. In 2007, EVERYTHING IS MISCELLANEOUS: THE POWER OF THE NEW DIGITAL DISORDER was published, looking at how the new ways of organizing stuff, freed from the limitations of the physical, are altering our institutions and ideas. I'm currently working on a book that looks back at the Information Age, not entirely with affection.

Over the past ten years, I've found myself happily pulled into discussions of the future of journalism, the use and abuse of conversational marketing, the

Internet as a tool to transform democracy (I had the inflated title of "Senior Internet Adviser" to the Howard Dean campaign, to which I was a volunteer consultant), Internet policy issues, and many more.

Over the past ten years, I've had an opportunity to speak with many companies around the world in many industries. I especially enjoy the chance to talk with companies that are starting up being startups. In fact, in 2000, Rick and I, along with my friend Paul English, started our own company, with Rick as CEO and WordOfMouth.com as the domain name. The idea I think was sound and quite Cluetrainy, but it required us to build a social networking system from scratch, before they were ready-to-hand. At least I had an excuse to come to Boulder frequently and spend time with Rick.

Since 2003, I've had the privilege of being a Fellow at Harvard's Berkman Center for Internet and Society (along with Doc), a source of constant stimulation, collegiality, and friendship. The Berkman Center has become a world hub of thought, research, and conversation about the effect of the Net. I could not be happier there.

Between the blogging (www.JohoTheBlog.com), the twittering, the speaking, the writing, the Berkman events, and serendipity of the inbox, I am feeling usefully overstimulated. Thank you, Internet!

INDEX